T0145219

Smart Innovation, Systems and Technologies

Volume 99

Series editors

Robert James Howlett, Bournemouth University and KES International,
Shoreham-by-sea, UK
e-mail: rjhowlett@kesinternational.org

Lakhmi C. Jain, University of Technology Sydney, Broadway, Australia;
University of Canberra, Canberra, Australia; KES International, UK
e-mail: jainlakhmi@gmail.com; jainlc2002@yahoo.co.uk

The Smart Innovation, Systems and Technologies book series encompasses the topics of knowledge, intelligence, innovation and sustainability. The aim of the series is to make available a platform for the publication of books on all aspects of single and multi-disciplinary research on these themes in order to make the latest results available in a readily-accessible form. Volumes on interdisciplinary research combining two or more of these areas is particularly sought.

The series covers systems and paradigms that employ knowledge and intelligence in a broad sense. Its scope is systems having embedded knowledge and intelligence, which may be applied to the solution of world problems in industry, the environment and the community. It also focusses on the knowledge-transfer methodologies and innovation strategies employed to make this happen effectively. The combination of intelligent systems tools and a broad range of applications introduces a need for a synergy of disciplines from science, technology, business and the humanities. The series will include conference proceedings, edited collections, monographs, handbooks, reference books, and other relevant types of book in areas of science and technology where smart systems and technologies can offer innovative solutions.

High quality content is an essential feature for all book proposals accepted for the series. It is expected that editors of all accepted volumes will ensure that contributions are subjected to an appropriate level of reviewing process and adhere to KES quality principles.

More information about this series at http://www.springer.com/series/8767

Vladimir L. Uskov · Robert J. Howlett
Lakhmi C. Jain · Ljubo Vlacic
Editors

Smart Education and e-Learning 2018

 Springer

Editors
Vladimir L. Uskov
Department of Computer Science and
    Information Systems, and InterLabs
    Research Institute
Bradley University
Peoria, IL
USA

Robert J. Howlett
Bournemouth University
Poole
UK

and

KES International
Shoreham-by-Sea
UK

Lakhmi C. Jain
Centre for Artificial Intelligence,
    Faculty of Engineering
    and Information Technology
University of Technology Sydney
Sydney, NSW, Australia

and

Faculty of Science, Technology
    and Mathematics
University of Canberra
Canberra, ACT, Australia

and

KES International
Shoreham-by-Sea, UK

Ljubo Vlacic
Griffith Sciences - Centres and Institutes
Griffith University
South Brisbane, QLD, Australia

ISSN 2190-3018   ISSN 2190-3026   (electronic)
Smart Innovation, Systems and Technologies
ISBN 978-3-030-06417-4   ISBN 978-3-319-92363-5   (eBook)
https://doi.org/10.1007/978-3-319-92363-5

Printed on acid-free paper

This Springer imprint is published by the registered company Springer International Publishing AG part of Springer Nature
The registered company address is: Gewerbestrasse 11, 6330 Cham, Switzerland

Preface

Smart education, smart e-learning, and smart universities are emerging and rapidly growing areas. They have a potential to transform existing teaching strategies, learning environments, and educational activities and technology in a classroom. Smart education and e-learning are focused at enabling instructors to develop new ways of achieving excellence in teaching in highly technological smart classrooms, and providing students with new opportunities to maximize their success and select the best options for their education, location, learning style, and mode of content delivery.

From June of 2014, the enthusiastic and visionary scholars, faculty, Ph.D. students, administrators, and practitioners have an excellent opportunity for a highly efficient and productive professional meeting—the annual international conference on Smart Education and Smart e-Learning (SEEL). The KES International professional association initiated SEEL conference as a major international forum for the presentation of innovative ideas, approaches, technologies, systems, findings and outcomes of research and design and development projects in the emerging areas of smart education, smart e-learning, smart pedagogy, smart analytics, applications of smart technology and smart systems in education and e-learning, smart classrooms, smart universities, and knowledge-based smart society.

The inaugural international KES conference on Smart Technology-based Education and Training (STET) has been held at Chania, Crete, Greece, on June 18–20, 2014. The 2nd international KES conference on Smart Education and Smart e-Learning took place in Sorrento, Italy, on June 17–19, 2015, the 3rd KES SEEL conference—in Puerto de la Cruz, Tenerife, Spain, on June 15–17, 2016, and the 4th KES SEEL conference—in Vilamoura, Portugal, on June 21–23, 2017.

The main topics of the SEEL international conference are grouped into several clusters and include but are not limited to:

- **Smart Education (SmE cluster):** conceptual frameworks for smart education; innovative smart teaching and learning technologies; best practices and case studies on smart education; smart pedagogy and innovative teaching and learning strategies; smart classroom; smart curriculum and courseware design and development; smart assessment and testing; smart learning and academic analytics; student/learner modeling; smart faculty modeling, faculty development and instructor's skills for smart education; university-wide smart systems for teaching, learning, research, management, safety, security; smart blended, distance, online and open education; partnerships, national and international initiatives and projects on smart education; economics of smart education;

- **Smart e-Learning (SmL cluster):** smart e-learning: concepts, strategies, and approaches; Massive Open Online Courses (MOOC); Small Personal Online Courses (SPOC); assessment and testing in smart e-learning; serious games-based smart e-learning; smart collaborative e-learning; adaptive e-learning; smart e-learning environments; courseware and open education repositories for smart e-learning; smart e-learning pedagogy, teaching and learning; smart e-learner modeling; smart e-learning management, academic analytics, and quality assurance; faculty development and instructor's skills for smart e-learning; research, design and development projects, best practices and case studies on smart e-learning; standards and policies in smart e-learning; social, cultural, and ethical dimensions of smart e-learning; economics of smart e-learning;

- **Smart Technology, Software and Hardware Systems for Smart Education and e-Learning (SmT cluster):** smart technology-enhanced teaching and learning; adaptation, sensing, inferring, self-learning, anticipation, and self-organization of smart learning environments; Internet of Things (IoT), cloud computing, RFID, ambient intelligence, and mobile wireless sensor networks applications in smart classrooms and smart universities; smartphones and smart devices in education; educational applications of smart technology and smart systems; mobility, security, access and control in smart learning environments; smart gamification; smart multimedia; smart mobility;

- **"From Smart Education to Smart Society" Continuum (SmS cluster):** smart school; applications of smart toys and games in education; smart university; smart campus; economics of smart universities; smart university's management and administration; smart office; smart company; smart house; smart living; smart health care; smart wealth; smart lifelong learning; smart city; national and international initiatives and projects; smart society;

- **"Smart University as a Hub for Students' Engagement into Virtual Business and Entrepreneurship" (SmB cluster):** entrepreneurship and innovation at university: student role and engagement; student engagement with virtual businesses and virtual companies; virtual teams and virtual team working (technology, models, ethics); university curricula for entrepreneurship and innovation (core and supplemental courses); new student goal—start his/her own business (instead of getting a job in a company); students and start-up companies (approaches, models, best practices and case studies).

One of the advantages of the SEEL conference is that it is organized in conjunction with several other Smart Digital Futures (SDF) high-quality conferences, including Agents and Multi-agent Systems: Technologies and Applications

(AMSTA), Intelligent Decision Technologies (IDT), Intelligent Interactive Multimedia Systems and Services (IIMSS), Innovation in Medicine and Healthcare (IMH), and Smart Transportation Systems (STS). This provides SEEL conference participants with unique opportunities to attend also AMSTA, IDT, IIMSS, IMH, and STS conferences' presentations, meet and collaborate with subject matter experts in those "smart" areas—fields that are conceptually close to SEEL areas.

This book contains the contributions presented at the 5th international KES conference on Smart Education and e-Learning (SEEL-2018), which took place at Gold Coast, Australia, on June 20–22, 2018. It contains peer-reviewed chapters that are grouped into several interconnected parts: Part 1—Smart Education: Systems and Technology, Part 2—Smart Pedagogy, Part 3—Smart Education: Case Studies and Research, and Part 4—Sustainable Learning Technologies: Smart Higher Education Futures.

We would like to thank many scholars who dedicated many efforts and time to make SEEL international conference a great success, namely: Dr. Farshad Badie (Denmark), Dr. Jeffrey P. Bakken (USA), Dr. Elena Barbera (Spain), Prof. Madhumita Bhattacharya (New Zealand), Dr. Janos Botzheim (Hungary), Dr. Claudio da Rocha Brito (Brazil), Dr. Dumitru Burdescu (Romania), Dr. Melany M. Ciampi (Brazil), Prof. Steven Coombs (UAE), Dr. Juan Manuel Dodero (Spain), Mr. Marc Fleetham (UK), Dr. Ekaterina Prasolova-Førland (Norway), Dr. Jean-Pierre Gerval (France), Dr. Foteini Grivokostopoulou (Greece), Dr. Karsten Henke (Germany), Dr. Alexander Ivannikov (Russia), Dr. Gara Miranda Valladares (Spain), Dr. Marina Lapyonok (Russia), Dr. Andrew Nafalski (Australia), Dr. Toshio Okamoto (Japan), Dr. Enn Õunapuu (Estonia), Dr. Mrutyunjaya Panda (India), Dr. Isidoros Perikos (Greece), Dr. Valeri Pougatchev (Jamaica), Dr. Luis Anido Rifón (Spain), Prof. Jerzy Rutkowski (Poland), Dr. Demetrios Sampson (Australia), Dr. Danguole Rutkauskiene (Lithuania), Dr. Adriana Burlea Schiopoiu (Romania), Dr. Ruxandra Stoean (Romania), Dr. Masanori Takagi (Japan), Dr. Wernhuar Tarng (Taiwan), Dr. Yoshimi Teshigawara (Japan), Prof. Toyohide Watanabe (Japan), Dr. Heinz-Dietrich Wuttke (Germany), and Dr. Larisa Zaiceva (Latvia).

We also are indebted to international collaborating organizations that made SEEL international conference possible, specifically: KES International (UK), InterLabs Research Institute, Bradley University (USA), Science and Education Research Council (COPEC), Institut Superieur de l'Electronique et du Numerique ISEN-Brest (France), Silesian University of Technology (Poland), Multimedia Apps D&R Center, University of Craiova (Romania), and World Council on System Engineering and Information Technology (WCSEIT).

We look forward to continuing the successful SEEL international conference, and we plan SEEL-2019 to be held at Malta on June 16–19, 2019.

It is our sincere hope that this book will serve as a useful source of valuable collection of knowledge from various research, design and development projects, useful information about current best practices and case studies, and provide a baseline of further progress and inspiration for research projects and advanced developments in Smart Education and Smart e-Learning areas.

June 2018

Prof. Vladimir L. Uskov, Ph.D. (USA)
Prof. Robert J. Howlett, Ph.D. (UK)
Prof. Lakhmi C. Jain, Ph.D. (Australia)
Prof. Ljubo Vlacic, Ph.D. (Australia)

Contents

Contents

Smart Education:
Systems and Technology

Smart Learning Analytics: Conceptual Modeling and Agile Engineering

Vladimir L. Uskov[1(✉)], Jeffrey P. Bakken[2], Ashok Shah[1],
Timothy Krock[1], Alexander Uskov[1], Jitendra Syamala[1],
and Rama Rachakonda[1]

[1] Department of Computer Science and Information Systems and InterLabs
Research Institute, Bradley University, Peoria, IL, USA
`uskov@fsmail.bradley.edu`
[2] The Graduate School, Bradley University, Peoria, IL, USA
`jbakken@fsmail.bradley.edu`

Abstract. Learning analytics focuses on collecting, cleaning, processing, visualization and analyzing teaching and learning related data or metrics from a variety of academic sources. Our vision for engineering of smart learning analytics – the next generation of systems and tools for learning analytics - is based on the concept that this technology should strongly support "smartness" levels of smart academic institutions such as adaptivity, sensing, inferring, anticipation, self-learning, and self-organization. This paper presents the up-to-date findings and outcomes of research, design and development project at the InterLabs Research Institute at Bradley University (Peoria, IL, U.S.A.) that is focused on conceptual modeling of smart learning analytics systems, including identification of goals, objectives, features and functions, main components, inputs and outputs, hierarchical and smartness levels, mathematical methods and algorithms for those systems. Agile software engineering approach has been used for a development of a series of software prototypes to verify the design and development process and validate the obtained outcomes for smart learning analytics systems.

Keywords: Smart Learning Analytics · Conceptual modeling
Smart Education

1 Introduction

Academic institution analytics (or, institutional analytics) is an emerging field of research, design, development and practice that is focused on the active use of institutional big data analysis. It is aimed to help academic stakeholders – administrators, faculty, students, IT staff, etc. – make well-justified decisions in every major business function of the academic institution, for example, student enrollment, student retention, student success, student academic performance, evaluation of a faculty, program academic review, etc.

In accordance with the Hanover Research (USA) report [1], "Analytics used by educational institutions can cover a broad range of types, data sources, and areas for

V. L. Uskov et al. (Eds.): KES SEEL-18 2018, SIST 99, pp. 3–16, 2019.
https://doi.org/10.1007/978-3-319-92363-5_1

implementation. The broad types or focus areas for analytics within the higher education sphere can be divided into two major segments: Institutional/Academic Analytics (AA) and Learning Analytics (LA). AA emphasizes performance of the university as a whole and tends to echo the frameworks, techniques, and purposes of Business Analytics. It may incorporate learning performance data, but aggregated at the institutional, regional, or national level to illustrate performance of the university. LA emphasizes the learning process, or the condition and performance of the individual learner. This type of analysis may be targeted toward the instructor and/or to the student himself or herself".

According to the EDUCAUSE (USA) reports [2, 3], the priorities for AA and LA at 245 surveyed U.S. universities are presented on Fig. 1 [2].

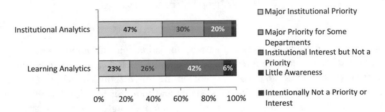

Fig. 1. The priorities for Academic Analytics and Learning Analytics at U.S. universities in 2015 [2]

"Only 23% of respondents … said learning analytics was a major institutional priority; another 26% identified it as a major priority for some departments or units, but not an institutional one. For 4 in 10 respondents, it was "an interest, but not a priority. By contrast, twice as many (47%) described institutional analytics as a major institutional priority, and an additional 30% called it a departmental one. It is important to note that interest and priority may align differently among different populations of institutional employees. Learning analytics may be reported as a higher priority among academic leaders than among IT professionals" [3].

Both AA and LA may use data from various hierarchical levels, for example, including levels of data related to (1) individual student, (2) individual faculty, (3) an academic course or a group of courses, (4) an institution's office or academic department, (5) a university's college and/or school, (6) a university as a whole, (7) a group of institutions in a region or state, and (8) institutions on a national level. LA and AA may run multiple types of related analysis for academic institutions; some examples are presented in Fig. 2 [1, 4–7].

In our previous works [8, 9] we reviewed and analyzed multiple publications regarding AA and LA, and, particularly, we identified and analyzed about 25 tools, systems and technical platforms that are currently in use by academic institutions to provide various types of analysis in AA and LA. For example, those systems include (1) commercial systems such as Google Analytics, Adobe Analytics, Clicky, Woopra, etc., (2) open source or free systems or systems developed by academic institutions – Piwik, Open Web Analytics, Klass Data, eAnalytics, Jpoll, Moodog, Equella, E2Coach, Signals, Sherpa, etc., and (3) platforms and/or plugins such as Blackboard

ANALYTICS CATEGORY	LEVEL OF ANALYSIS	TYPE OF ANALYSIS
Learning Analytics	Course Level	■ Social Network Analysis ■ Conceptual Development Analysis ■ Discourse Analysis ■ Personalized Curriculum ■ Student Performance Assessment
	Student Level	■ Degree Audit ■ Performance Assessment ■ Predictive Performance Analysis/Early Warning Systems ■ Automated Advising and Coaching
	Departmental Level	■ Early Warning/Predictive Modelling
Institutional Analytics	Instructor Level	■ Teacher Effectiveness ■ Financial Contributions
	Student or Student Body Level	■ Enrolment Profiling and Predictive Analysis ■ Lifetime Value/Booster Effectiveness ■ Advocacy ■ Post-Educational Employment Analysis ■ Subject or Course Selection Recommendations
	Institutional Level	■ Admissions Analysis ■ Institutional Performance/Efficiency ■ Retention/Attrition Trends
	Public Level	■ Comparison with Other Institutions

Fig. 2. Examples of types of data analysis and levels of analysis employed by higher education institutions [4–7].

Intelligence, Blackboard Predict, Blackboard X-Ray Learning Analytics, Moodle Analytics plugins, SEAtS Learning Analytics, and others.

One of the outcomes of our analysis shows that existing systems for AA and LA in most of the applications for AA and LA either already complement, or, are planned to be integrated in the future and complement each other. For example, an LA system on the level of the individual student (for example, placing a student in a correct course in a program of study) may need information from an AA system – Admission Management or Enrollment Management system - on an institutional level. On the other hand, an institutional Student Retention Management system in AA cluster may require information from an LA system on student or departmental levels (for example, data on student academic performance in math courses or writing intensive courses). This is the reason that below we will discuss only systems related to LA – because most of our proposed conceptual solutions for LA systems will be valid for AA systems as well.

Unfortunately, the analyzed publications in AA and LA fields do not provide detailed information about the next generation of systems for LA – Smart Learning Analytics (SLA). Those systems are aimed to strongly support the concepts of Smart University (SmU), Smart Education (SmE), Smart Pedagogy (SmP) and Smart Classroom (SmC), including their smartness levels such as (1) adaptivity, (2) sensing, (3) inferring, (4) anticipation, (5) self-learning, and (6) self-organization [8–11].

Additionally, only a few publications (for example, [12]) described initial approaches to apply the well-thought and powerful Gartner Analytics Ascendancy

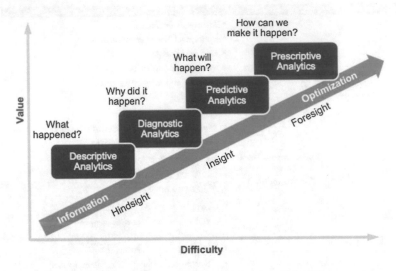

Fig. 3. Gartner analytics ascendancy model [13]

Model (Fig. 3) [13] to LA systems with multiple hierarchical levels – student, faculty, course, and departmental levels. The Gartner model describes consecutive hierarchical levels on analytics – descriptive, diagnostic, predictive and prescriptive analytics. We believe this model creates a very solid foundation for the conceptual modeling and engineering (e.g. analysis, design development) of SLA systems.

2 Project Goal and Objectives

The overall goal of the on-going multi-aspect research, design and development project at the InterLabs Research Institute at Bradley University (IL, USA) is to focus on conceptual modeling and engineering of Smart Learning Analytics systems for Smart Education.

In order to achieve this goal, the project team selected the following objectives:

- identify types of data and/or metrics to be collected and processed by AA and LA systems (our research findings for this objective are presented in [8]);
- analyze existing commercial and open source systems and technical platforms and plugins for AA and LA (our research findings for this objective are presented in [8, 9]);
- identify requirements for functionality of SLA systems (our research findings for this objective are presented in [9]);
- develop a conceptual model of a SLA system, including system's goals, main components, features, hierarchical levels, inputs, outputs, metrics, mathematical methods and algorithms, types of data analysis, and engines;
- develop a structural model of a SLA system, including main system's components and links between them;

- analyze various algorithms for predictive analytics in a SLA system, and
- apply agile software engineering for a prototyping of a SLA system.

A summary of up-to-date project findings and outcomes is presented below.

3 Smart Learning Analytics: Conceptual and Architectural Models

The conceptual model of a SLA system – *CM-SLA* - can be described as follows.

Definition. Smart Learning Analytics system is described as n-tuple of the following elements:

$$CM - SLA = \ <\{G, OBJ, FF, COMP, LINKS, IN, OUT, INT, LEV, EXT, LIM\} >$$

$$(1)$$

where:

{G}, {OBJ}, and {FF} – sets of goals, objectives and features/functions of SLA systems;

{COMP} – main interrelated components of an SLA system;

{LINKS} – links between interrelated components in SLA systems; they should clearly show what system's features and functions are available for which stakeholders;

{IN} and {OUT} – sets of inputs to and outputs from an SLA system;

{INT} – interfaces, i.e. a convention about legal types of data and data exchange protocols to be used in communications between SLA system's components and the environment;

{ALG} – a set of multiple mathematical methods and algorithms to be used by the system;

{LEV} – various levels of SLA system's architecture and operation;

{EXT} – external collaborative systems;

{LIM} – limits, restrictions or constrains of SLA systems

Most of sets in model (1) may have multiple subsets; for example, the {COMP} set of main systems' components can be described by the following model:

$$\{COMP\} = \ <\{STA, TOOLS, HW\}$$

$$(2)$$

where {STA}, {TOOLS} and {HW} are sets of SLA system's stakeholders, internal tools/engines and hardware.

The descriptions of several of the most important sets of models (1) and (2) are given below.

Goal. The goal of a SLA system is to provide LA stakeholders with analytics-based features and functionality to perform main business functions in academia – teaching, learning, tutoring, mentoring, research and administrative services – in an optimal way and with the highest possible quality.

Objectives. The objectives of SLA systems include but are not limited to: (1) student success control; (2) student academic progress control; (3) identification of *at-risk* students; (4) early warning (alerts); (5) personalization; (6) analysis and prediction; (7) recommendations; (8) interventions; (9) quality assurance; and (10) performance assessment.

Features/Functions. The main features and functions of SLA systems include but are not limited to: (1) adaptation; (2) sensing (i.e. getting data from sensors); (3) inferring (logical conclusions); (4) system's self-description and self-learning; (5) monitoring and anticipation; (6) system's self-organization and self-optimization; (7) student self-assessment and self-control of learning; (8) automated tutoring/coaching/mentoring/ advising; (9) assessment of faculty teaching and performance; (10) instructional management; (11) various types of data analysis for descriptive, diagnostic, predictive and prescriptive analytics, (12) logfile analysis, and other functions and types of data analysis; some of them are presented on Fig. 2 above.

Inputs. The inputs to SLA systems may be generated by a great variety of systems, websites, social networks, search engines, web analytics, etc. in a form of datasets, reports, documents, student transcripts, audits of student program of study, surveys, evaluations, blogs, posts, user logs, etc.

The main sources of data for SLA systems include but are not limited to: (1) student profile data such as lists of student's current courses and courses taken so far, remained courses in program of study, majors, minors, concentrations, GPA, etc.; (2) student academic performance data such as scores on learning assignments, tests, quizzes, labs, exams, etc. in academic course(s); (3) student learning-related activities' data such as a frequency of logs to learning management systems (LMS) and or websites of online courses, time spent to watch video lectures or participate in online discussions, the number and quality of posted questions or statements in discussions forums, etc.; (4) course syllabi, curriculum, program of study; (5) academic department related data such as admission criteria, offered academic programs, requirements to graduation, lab or technological fees, etc.; (6) college and/or university related data such as constraints on credit hours per semester – min and max – to be taken by a student in one semester, constrains on number of courses to be taken in summer sessions, tuition fees, student-to-faculty ratios; max enrollment; and other types of inputs.

Stakeholders. The stakeholders of SLA systems are (1) students, (2) faculty/ instructors, (3) school/college/university administrators, (4) IT/network/security administrators, engineers and professional staff, etc.

Outputs. The outputs from SLA systems with analytics-based outcomes may have different forms such as (1) reports with visualized data, (2) documents with predictions, (3) lists of recommendations for intervention, (4) datasets, etc.

Functional Tools (Internal Tools). A set of system's internal software tools to provide and support system's objectives and features and functions may include but are not limited to (1) adaptation engine, (2) sensing/metrics/measurement engine,

(3) inferring engine, (4) self-learning engine, (5) self-optimization engine, (6) descriptive analytics engine, 7) diagnostic analytics engine, (8) predictive analytics engine, (9) prescriptive analytics engine, (10) intervention or recommendations engine, (11) feedback engine, and (12) Web analytics engine, and other tools.

Methods and Algorithms. A set of multiple mathematical methods and algorithms to be used in SLA systems for a) descriptive, diagnostic, predictive and prescriptive analytics and b) various types of data analysis, including statistical analysis, regressions analysis, parsing, etc., for example, (1) machine learning algorithms, (2) data mining algorithms, (3) text-mining algorithms, (4) empirical rules, (5) search algorithms and engines, and other types of algorithms. In its own turn, a set of algorithms for regression analysis only may include but is not limited to algorithms for (1) linear regression, (2) logistic regression, (3) probit regression, (4) multinomial logistic regression, (5) non-linear regression, (6) non-parametric regression, (7) robust regression, (8) stepwise regression, and other types of algorithms.

Collaborative Systems (External Systems). SLA systems may exchange data with various types of systems such as (1) learning management system (LMS), (2) enrollment management system, (3) retention management system, (4) student degree planning system, (5) course management system or course website, (6) document management system, (7) curriculum management tool, (8) Web analytics tools, (9) faculty teaching quality/performance (or, student assessment of faculty teaching and academic courses), (10) tools for automatic tutor/instructor reports from writing intensive centers or math tutoring center, and multiple other systems.

Levels. Levels of SLA systems may include (1) hierarchical levels of its scope of operation (e.g. levels of student, faculty, course, department, or institution), (2) smartness levels of SLA systems, (3) levels of analytics provided (descriptive analytics, diagnostic analytics, predictive analytics and prescriptive analytics), and (4) levels of quality or maturity levels (initial, managed, defined, quantitatively managed, and optimizing levels).

Our approach to engineer the SLA system is based on the idea that SLA system – as a smart system – should provide the following "smartness" features: (1) adaptation, (2) sensing (or, awareness), (3) inferring (or, logical reasoning), (4) self-learning (a note: self-description and self-discovery features are a part of self-learning), (5) anticipation, and (6) self-organization (a note: self-optimization, self-protection, self-matchmaking, and self-healing are a part of self-organization).

Hardware. A set of hardware, servers, routers, equipment, electronics, technologies, etc. to support quality and secure operation of SLA systems.

Limits and Constraints. The SLA system may have constraints in terms of its functions, types and openness of data used and recommendations generated; those limits can deal with (1) privacy, (2) ethics, (3) institutional policies, (4) success/failure metrics and measures, and (5) transparency of data and recommendations, and others.

The SLA conceptual models (1) and (2) enabled us to develop a structural model of an SLA system - it presents the main SLA components and links between them (Fig. 4).

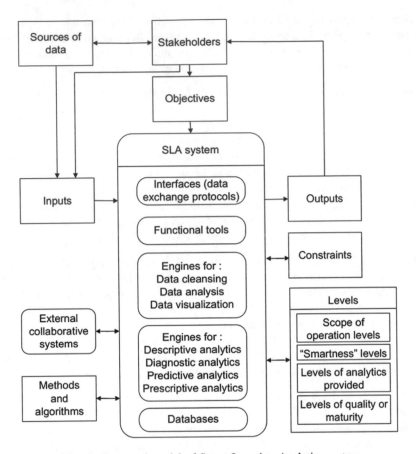

Fig. 4. Structural model of Smart Learning Analytics system

We would like to emphasize that it includes (1) corresponding engines for all four types of analytics of the Gartner Analytics Ascendancy Model, (2) hierarchical processing levels for various levels of LA (institutional, departmental, course, faculty, student), and (3) all six designated "smartness" levels as crucial and mandatory components of a SLA system.

4   Analysis and Testing of Predictive Algorithms for SLA

One of the unique features of the designed SLA system is its ability to perform predictive analytics as a part of SLA, for example, a prediction of students' final scores (and, as a result, final grades) in academic courses. This part of the SLA system follows the Gartner Analytics Ascendancy Model. A list of possible predictive algorithms that may potentially be implemented in SLA and types of their applications in SLA are presented in Table 1.

Table 1. A list of algorithms for predictive analytics in SLA system

Algorithm	Algorithm's main idea	Potential applications in SLA system
Linear Regression	Method for modeling linear relationships between a continuous dependent variable and one or more independent variable(s)	• Final score/grade prediction • Inter-assignment relation exploration
k-Nearest Neighbors Regression	It estimates output values of new data based on an average of the most similar past data points	• Final score/grade prediction • Comparison of student's performance to only "similar" students
Ensemble Methods	Combination of algorithms designed to compensate for various weaknesses between individual machine learning models	• Predict the behavior of students with different types of behavior (e.g. consistent students vs. inconsistent students)
Logistic Regression	Method for modeling relationships between a binary independent variable and one or more independent variables	• Locate indicators/sensors that a student is an *at-risk* student • Predict probability of failure given past data • Identify learning assignments that are crucial to student success in a course
Bayesian Networks	Directed probabilistic graphical model that applies a Bayesian understanding of probability to make predictions	• Course sequencing recommendations • Final score/grade prediction
Naïve Bayes Classifiers	Classifier that assumes features are independent so that Bayes' theorem can be applied "naively"	• Course sequencing recommendations • Identification of *at-risk* students (e.g. students with poor academic performance or attitude to learning)
Decision Trees	Predictive model of decisions and their possible consequences in the form of a tree	• Course Sequencing Recommendations • Critical assignment analysis (e.g. What assignments are critical for development of student's analytical, technical, management skills?)
Neural Networks	Predictive model in the form of a graph of nodes emulating the behavior of neurons in the brain. It is trained by altering graph weights to minimize simulation output error. (A note: usually it works with large amounts of data, or, big data	• Identify student's struggle with academic course(s) • Identify cheating • Evaluate student efforts

(*continued*)

Table 1. (*continued*)

Algorithm	Algorithm's main idea	Potential applications in SLA system
k-Nearest Neighbors Classifiers	Identification of the closest existing data to classify incoming data points based on a majority vote system	• Classify students into groups based on individual capabilities/skills (e.g. for a development of student's individual learning trajectory) • Calculate likelihood of student retention
Support Vector Machines	It is aimed to create a decision boundary (line or n-dimensional hyperplane) that best separates the data; this decision boundary can then be applied to new data	• Final score/grade prediction • *At-risk* student identification and classification • Identification of cheating

We tested the Linear Regression and k-Nearest Neighbors Regression predictive algorithms for the purposes of SLA. The goal of testing experiments was to identify the accuracy of the algorithms in terms of prediction of final student scores in a course. At the end of experiments, we were able to compare actual scores with predictions made by those two predictive algorithms.

(A disclaimer: All data used in testing experiments of this research project and presented in this paper are anonymized, e.g. neither names of actual students nor students' correspondence to those data were disclosed at any point of this research. Additionally, the input data presented in this paper have been proportionately modified in such a way that they are different from actual scores by any current or past student in that course. At the same time, the input data used in our experiments adequately reflect learning outcomes of students in that course).

The training dataset contained academic performance data - numeric scores (each on 100% basis) obtained by students for 8 course learning assignments in one of the academic courses for a time period between the beginning of the course and midterm (approximately in the middle of the course). We also provided algorithms with data about academic performance of about 60 students in this course in the past. Particularly, this data included student academic performance in the remaining 6 course learning assignments for a time period between the midterm and the final exam. This was done in order to identify scores for all course learning assignments and behavior of the "average" student score in this course in the past. The outcomes of testing of both designated predictive algorithms are presented in Table 2 for only 2 learning assignments in the 2nd half of the course–the course project and the final exam, and the total score obtained in that course.

The data in Table 2 presents the numeric outcomes of accuracy of our implementations of Linear Regression and the k-Nearest Neighbors regression algorithms. Particularly, we can make the following conclusions from our testing experiments:

(1) the k-Nearest Neighbors algorithm outperforms the Linear Regression in predicting the grades for particular course learning assignments;

Table 2. Quality of two tested algorithms for a prediction of final student score in a course

Algorithm	Course project (pts)	Error (%)	Final exam (pts)	Error (%)	Total score in a course (pts)	Error (%)
Linear Regression Output						
Max pts	100.00		100.00		500.00	
	89.39	1.44	94.47	1.53	475.25	0.81
	83.00	2.50	85.30	9.70	424.97	0.75
	95.73	1.69	96.26	3.74	506.91	2.27
	62.18	32.15	86.04	1.96	381.22	7.36
	84.58	6.84	91.17	1.83	451.91	5.64
	84.14	6.86	92.70	2.70	467.60	3.14
	82.22	10.76	91.93	9.93	451.83	8.76
Average error		**8.89%**		**4.48%**		**4.11%**
k-Nearest Neighbors Regression Output						
Max pts	100.00		100.00		500.00	
	95.57	4.74	92.70	3.30	480.58	1.88
	89.56	9.06	89.60	5.40	453.40	6.44
	92.41	1.63	92.30	7.70	478.58	7.94
	94.12	0.21	88.90	0.90	469.43	10.28
	93.69	2.27	90.50	2.50	473.74	1.27
	92.51	1.51	89.80	0.20	473.74	1.91
	91.96	20.50	88.50	6.50	466.38	11.67
Average error		**5.70%**		**3.79%**		**5.91%**

(2) the Linear Regression algorithm predicted better final scores of students;
(3) each algorithm has outliers that contribute significantly to the average error; however the outliers are unique to each algorithm. For this reason, an adaptive algorithm may be able to improve upon the performance of designated two individual algorithms.

5 SLA: Agile Engineering and Prototyping

Using the agile software engineering methodology: (a) "most important features/functions first" and (b) "new working prototype every 2 weeks", we consecutively developed several prototypes of the SLA system – the InterLabs SLA system. This system should eventually include 100% of functionality of the SLA as presented in [8]; it is also planned to implement most of types of analysis presented on Fig. 2. The examples of Student Dashboard, Faculty Dashboard and Administrator Dashboard of the current InterLabs SLA system – Sprint # 9 are presented on Figs. 5, 6 and 7 accordingly.

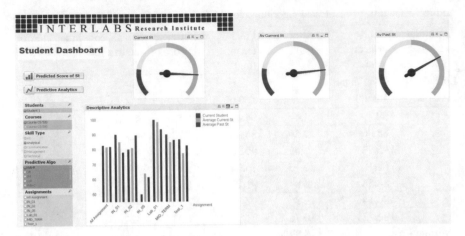

Fig. 5. The InterLabs Smart Learning Analytics system: Student Dashboard (Outcome data are shown for Descriptive Analytics outcomes for "start date – midterm date" time period with scores for all learning assignments by (a) a selected current student, (b) "average" current student, and (c) "average" past student in a selected course).

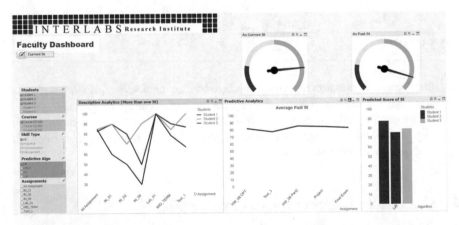

Fig. 6. The InterLabs Smart Learning Analytics system: Faculty Dashboard (Outcome data are shown for Descriptive Analytics from start date to midterm date with scores for all learning assignments by (a) several selected current students. Additionally, the Predictive Analytics outcomes - (b) score of "average" past student after midterm, and (c) predicted final scores of selected current students in that course, are shown as well).

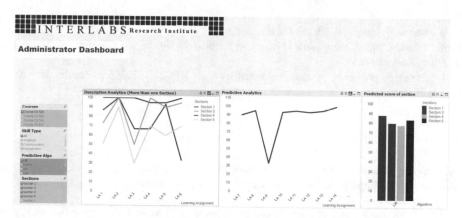

Fig. 7. The InterLabs Smart Learning Analytics system: Administrator Dashboard (Outcome data are shown for Descriptive Analytics from start date to midterm date with average scores obtained by students for (a) several sections of the same course. Additionally, the Predictive Analytics outcomes, e.g. (b) score of "average" past student after midterm, and (c) predicted final scores of "average" current student in all selected sections of that course, are shown as well).

6   Conclusions and Next Steps

Conclusions. The performed research and analysis, and obtained findings and outcomes enabled us to make the following conclusions:

(1) SLA systems will soon play a crucial role in academic institutions in terms of (a) student success, (b) improvement of student retention rate, (c) student course-level performance, and (d) reduction of degree completion time by students.

(2) The main research findings and deliverables of this project include (a) the developed conceptual models and identified and described main components (or, sets) of a SLA system (Sect. 3), (b) the obtained outcomes of benchmarking of two predictive algorithms for the purposes of SLA (Sect. 4), and (c) a developed prototype of a SLA system using agile software engineering methodology (Sect. 5).

Next Steps. Based on obtained research/design/development findings and outcomes, the next step in this research project is to involve various types of stakeholders into a testing of various features and functions of the InterLabs SLA system and further prototyping to ensure quality of proposed conceptual and software solutions for the InterLabs SLA system.

Acknowledgements. The authors would like to thank Mr. Cade McPartlin, Mr. Nicholas Hancher and Ms. Lavanya Aluri - the research associates of the InterLabs Research Institute and graduate and undergraduate students of the Department of Computer Science and Information Systems (CS&IS) at Bradley University - for their valuable contributions to this research, design and development project.

We also would like to thank Dr. Chris Jones, Professor and LAS Dean, and Dr. Steven Dolins, Professor and Chair of the CS&IS Department for their strong support of our research activities in Smart Education area.

This research project is partially supported by grant REC # 1326809 at Bradley University (2015–2018).

References

1. Learning Analytics For Tracking Student Progress. Hanover Research, November 2016. https://www.imperial.edu/research-planning/7932-learning-analytics-for-tracking-student-progress/file
2. Analytics Landscape: A Comparison of Institutional and Learning Analytics in Higher Education. EDUCAUSE, April 2016. https://library.educause.edu/~/media/files/library/2016/4/eig1504.pdf
3. Arroway, P., Morgan, G., O'Keefe, M., Yanosky, R.: Learning Analytics in Higher Education. Research report. Louisville, CO: ECAR, March 2016. https://library.educause.edu/resources/2016/2/learning-analytics-in-higher-education
4. Clow, D.: The learning analytics cycle: closing the loop effectively. In: Proceedings of the 2nd International Conference on Learning Analytics and Knowledge, LAK 2012, pp. 134–135. ACM, New York (2012)
5. Schmarzo, B.: What Universities Can Learn from Big Data – Higher Education Analytics. InFocus Blog|Dell EMC Services, 2 July 2014. https://infocus.emc.com/william_schmarzo/what-universities-can-learn-from-big-data-higher-education-analytics/
6. Bienkowski, M., Feng, M., Means, B.: Enhancing Teaching and Learning Through Educational Data Mining and Learning Analytics: An Issue Brief. U.S. Department of Education, October 2012. http://tech.ed.gov/wp-content/uploads/2014/03/edm-la-brief.pdf
7. Uskov, V.L. Bakken, J.P., et al.: Building smart learning analytics system for Smart University. In: Uskov, V.L., Howlett, R.J., Jain, L.C. (eds.) Smart Education and e-Learning 2017, pp. 191–204. Springer, June 2017. https://doi.org/10.1007/978-3-319-59451-4, ISBN 978-3-319-59450-7
8. Uskov, V.L., Bakken, J.P., Howlett, R.J., Jain, L.C. (eds.): Smart Universities: Concepts, Systems, and Technologies, 421 p. Springer (2018). ISBN 978-3-319-59453-8
9. Uskov, V.L., et al.: Smart pedagogy for Smart Universities. In: Uskov, V.L., Howlett, R.J., Jain, L.C. (eds.) Smart Education and e-Learning 2017, pp. 3–16. Springer (2017). https://doi.org/10.1007/978-3-319-59451-4, ISBN 978-3-319-59450-7
10. Burlea Schiopoiu, A., Burdescu, D.D.: The development of the critical thinking as strategy for transforming a Traditional University into a Smart University. In: Uskov, V., Howlett, R., Jain, L. (eds.) Smart Education and e-Learning 2017, Smart Innovation, Systems and Technology, vol. 75, pp. 67–74. Springer International Publishing, Cham (2017)
11. Uskov, V.L., Bakken, J.P., et al.: Learning analytics based smart pedagogy: student feedback. In: Uskov, V., Howlett, R., Jain, L. (eds.) Smart Education and e-Learning 2018, Smart Innovation, Systems and Technology. Springer, Cham (in print, this volume)
12. Boyer, A., Bonnin, G.: Higher Education and the Revolution of Learning Analytics. International Council for Open and Distance Education. https://icde.memberclicks.net/assets/Members_area, ISBN 978-82-93172-38-3
13. Elliot, T.: #GartnerBI: Analytics Moves To The Core. https://timoelliott.com/blog/2013/02/gartnerbi-emea-2013-part-1-analytics-moves-to-the-core.html

GOLDi-Lab as a Service – Next Step of Evolution

Karsten Henke[1], Heinz-Dietrich Wuttke[1(✉)], René Hutschenreuter[1], and Aleander Kist[2]

[1] Technische Universität Ilmenau, Ilmenau Ehrenbergstr. 29, Germany
{karsten.henke,dieter.wuttke}@tu-ilmenau.de
[2] University of Queensland, Toowoomba, Australia

Abstract. The GOLDi-Lab is a cloud-based grid of online labs for teaching design and verification of digital control systems in hardware and software. The paper presents a concept of enhancing this lab with IoT- features to make it more flexible and to prepare learners for the Internet of Things (IoT).

Keywords: Lab as a service · Cloud services · Smart e-learning
IoT

1 Introduction

As mentioned in [1], smart technologies "have great potential, and may strongly support and significantly benefit the next generation on advanced technology-based learning - Smart Learning." Actually, our remote lab is in the third generation of development [2]. In its implementation, it uses modern web technologies but it is realized more or less as a monolithic hardware-software system with little documentation. Using parts of it as a service in a way, described e.g. in [3], is actually impossible, but important in a future Internet of Things environment.

To overcome this gap, we will develop the next lab generation as a smart lab. This concept we will discuss in the paper. The rest of the paper has the following structure: Sect. 2 discusses the state of the art in developing smart labs. Section 3 concludes the actual status of the GOLDi-Lab and gives an overview about its architecture. Section 4 discusses a possible modularization of the lab to define and provide separate parts of the lab as services and Sect. 5 gives an example of an ongoing project with an Australian university. Section 6 concludes the paper.

2 Related Works

As we see in a recherché, the term "Smart Laboratory" is widely used for industrial labs for chemical or medical analysis. In the field of education the SMART Technologies company has claimed the term "SMART" for their products like SMART board™, SMART notebook™ and the like and also SMART lab™. In that context, SMART lab™ is a software and framework for developing virtual educational games.

© Springer International Publishing AG, part of Springer Nature 2019
V. L. Uskov et al. (Eds.): KES SEEL-18 2018, SIST 99, pp. 17–26, 2019.
https://doi.org/10.1007/978-3-319-92363-5_2

We will use the term "smart lab" in relation to smart devices and the Internet of Things.

In [4] the authors characterize a smart device by four main abilities:

It should be able to

- detect hazards, such as fires and accidental spill of chemical and bacteria;
- monitor long-term health of staffs and students, as well as conditions of equipment and the environment;
- track the existence of equipment for maintaining safety and security; and
- regulate the environment to reduce consumption of power and other resources without deteriorating efficiency of research activities.

Therefore the authors suggest a cyber-physical system, equipped with some environmental sensors that monitors the conditions in the lab. That way it is able to detect out of range situations like high temperature, vibrations, unexpected noise, and the like.

Thomson suggested in [5] that smart devices need one or all of the following capabilities: communication, sensing and actuating, reasoning and learning, identity and kind, memory and status tracking. Salzmann et al. extended these properties tailored to remote labs [6, 7]. In doing this, they specify the RL interface to the Internet as well as the internal functionalities.

3   Current Situation

The architecture of the GOLDi-lab relays on a cloud-based client server architecture with a lab-server connected to GOLDi-lab devices like control units (CUs) such as FPGAs or microcontrollers and controllable objects (COs) e.g. elevator, production cell, 3-axis-portal, and the like. Actually, ten GOLDi-lab-instances in four countries have access to this cloud.

The students on their client PC or mobile device must create control algorithms in different languages (e.g. VHDL, C) using the cloud-based design environment and upload them into the execution environment (Execution control panel, ECP) of the GOLDI system. The ECP delivers the files to the appropriate CU. While executing the student's control algorithm a validation unit named PSPU (physical system protection unit) protects the COs against malfunctions, caused by wrong student's design, by watching all I/O signals to/from the COs [3].

A Finite State Machine (FSM) as basis for a hardware or software design defines the control process logic (CPL). At the client side, there is a common user interface, showing the webcam image and the actuator/sensor values as well as buttons to control the experiment (start, stop etc.). The design environment helps producing the appropriate files for the CUs and supports uploading them to the CUs via the lab-server. The PSPU checks the validity of each value before it reached the appropriate actuators to control the experiment's CO.

Figure 1 shows an overview of the GOLDI-lab structure. Each instance of the GOLDi-lab has his own lab-server and lab devices but uses the same cloud services like development tools (GIFT, BEAST, ECP upload service etc.) and user management

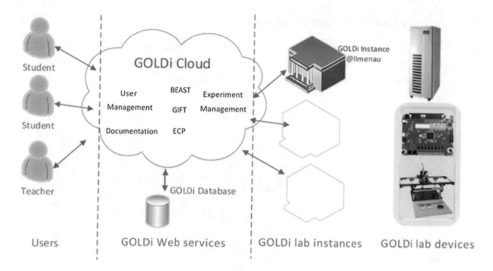

Fig. 1. GOLDI-lab Overview

(login, booking service and the like. The maintenance and management of the lab devices is also part of the cloud services.

The next subsections describe the GOLDI-lab hardware and cloud components more detailed to analyze, which parts are subject for extending to introduce smart lab properties.

3.1 Cloud Services

Available GOLDi Servers (each corresponding to a partner remote lab) are registered in the GOLDi cloud.

Each GOLDi user communicates with the GOLDi cloud to access the following GOLDi Web services:

- Documentation: GOLDi user manual about CUs, COs, experiment tasks, example solutions and the like
- User Management: login at any lab-instance via single sign in,
- ECP (Experiment Control Panel): User interface to perform remote experiments and provide the communication with the GOLDi lab equipment Partner Lab Server,
- GIFT (Graphical Interactive Finite State Machine Tool): to develop Finite State Machine based designs [8],
- BEAST (Block diagram Editing and Simulation Tool): to develop block diagram based digital circuit designs [9],
- Experiment Management: Experiment configuration, booking and pre-planning.

All GOLDi-lab instances use these services. That way, all partners have access to the actual version and updates have to be implemented only ones.

3.2   GOLDI-Lab Hardware Components

The GOLDi-lab hardware structure, shown in Fig. 2, bases on extensible grid, which guarantees a reliable, flexible as well as robust configuration of the CU/CO assignment. For a more detailed description of this grid concept, see [10] and [11].

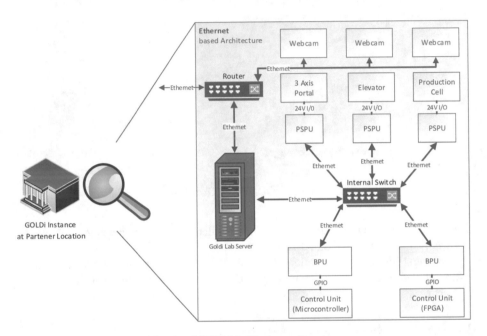

Fig. 2. GOLDi-lab Hardware Structure

The server side infrastructure (remote lab) consists of three parts:

- An internal **Remote Lab Bus**, realized as a common Ethernet based solution. This results in the lab server being able to communicate directly with all connected lab devices.
- a **Bus Protection Unit** (BPU) to interface the selected control unit with the remote lab bus and to protect the remote lab bus from misuse and damaging as well as.
- a **Physical System Protection Unit** (PSPU), which protects the COs against deliberate damage or accidentally wrong control commands and which offers different access and control mechanisms. Electrically, it performs the translation between the communication bus and 24 V I/O of the COs.

The remote lab server, which also handles and distributes the video streams generated by the web-cams realizes the interconnection between the CU and the selected CO during an experiment session as well as the connection to the cloud services.

Concerning the analysis done in [12], the GOLDi Laboratory is characterized by the following properties:

The lab has a very good software and hardware flexibility, offers share-ability of experiments with virtual and real COs and asses the user`s inputs in a very good manner with a feedback message as soon as the students' algorithm sends a wrong value to an actuator.

Figure 3 illustrates these facts. It also shows that there is a lack of support provided with the actual implementation of the GOLDi-lab.

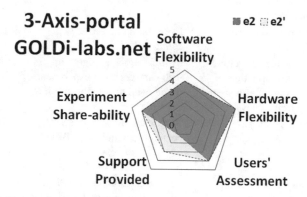

Fig. 3. SHASS Chart for 3-axis-portal experiment (*e2*) and its possible improvements (*e2'*) with respect to GOLDi-labs.net [12]

By introducing the smart device paradigm to the GOLDi-lab, we want to overcome with this identified deficit to open possibilities for renting parts of the lab as services [4].

Therefore we will compare the properties of smart lab, found in the literature, with the actual status of the GOLDi-lab in the next chapter.

4   Improvements of the GOLDi-Lab

Keeping in mind the above-mentioned requirements of smart laboratories, we compare the actual status of the GOLDi-labs with the requirements of smart devices and laboratories. Table 1 shows our findings.

The compare shows the strengths and weaknesses of the GOLDi-lab in terms of smart learning.

5   Example

An actual project with the University of Queensland aims to connect a special control unit, a programmable logic controller (PLC) located at that university, with the GOLDi-lab infrastructure [13].

RALfie was initially developed as an environment to easily integrate remotely accessible experiments for school students, and is presently being used to provide access to remote learning activities for university students. This environment provides

Table 1. Smart Lab properties and the GOLDi-lab

Reference	Smart lab	Actual GOLDi-implementation	Extension of GOLDI
[1, 4]	Detect hazards Self-protection	PSPU detects wrong signals to actuators	Exists
[1, 4]	Monitor conditions of equipment and the environment	Physical system protection unit per experiment	Exists
[4]	Track the existence of equipment	Event based detection of connected devices	Exists
[1, 4]	Regulate the environment to reduce consumption of power and other resources Energy efficiency	Switch on/off the light and web-cam per used experiment, on demand generation of video streams	Exists
[5]	Communicating	Sensor calculation by PSPU, actuator generation by BPU	D2D Communication possible due to shared medium
[5]	Sensing and actuating,	Sensor calculation by PSPU, actuator generation by BPU	Exists
[5]	Reasoning and learning,	Not available	
[1, 5]	Identity and kind, Self-description	Only ID, further information available via data base	Possible
[5]	Memory and status tracking	Tracking the users trials but without an evaluation	Exists
[1]	Self-optimization	Not available	Exists
[1]	Self-healing	Reset and initialization on demand and on errors	Exists
[1]	Self-discovery	Instantiation, remote CU (see chapter 5 in this paper)	In development
[1]	Self-matchmaking	Interface description available, devices can be used for any purpose	Exists
[1]	Compatibility	Supported Technologies: Ethernet/CAN/WIFI	Exists
[1]	Contextual awareness	Partly implemented for Learning Analysis	Further developments required
[1]	Dimension-ability	Real and virtual COs can be added seamlessly	Exists
[1]	Multimodal HCI	Not available	Exists
[1]	Usability	Available tools with some not intuitive requirements	To be improved
[1]	Connectivity	Web-interface based on HTML 5	Exists

mediated and authenticated access to remote laboratory equipment. The system separates hardware experiments from their corresponding activities: i.e. a hardware rig can have multiple different activities associated with it. Activities can use different control inter-faces to the lab. It supports custom web-based, Snap!$_1$-based or RDP-based user client interfaces. Common to all is access to live web camera streaming feeds from the experiment. The focus here is on the RDP-based interface.

The RALfie system supports authentication and access control. Once users are authenticated for a particular activity, the experiments can be accessed directly via specific URLs. For example, `cam1-pn.ralfie.net` enables direct access the video feed of the experiment and `panel-pn.ralfie.net` allows access to the virtual panel. The user sees the activity page that shows the experiment client user interface (here using RDP) and provides access to additional information, such as video stream feeds and virtual experiment controls via popup panels

Figure 4 shows an overview of the RALfie PLC Architecture located at the University of Queensland. On the right hand side, the physical lift model is shown. The sensors and actuators of the model are connected to the micro controller unit (MCU), which implements protection and interface functionality. The physical system protection unit ensures that commands sent to the physical model are within given parameters and tolerances and will not damage the physical model. The Virtual Inter-face Unit (VIU) makes the physical model inputs and outputs available via virtual panel. This includes the call buttons for the lift, for example. These are rendered as a web page.

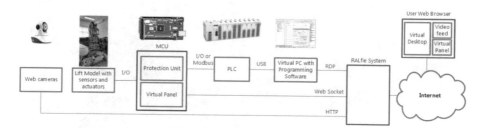

Fig. 4. RALfie PLC Architecture

The CU, i.e. the PLC, is connected to the protection unit via I/O lines. Alternatively, this connection can also be achieved via Modbus. In turn, the PLC is connected via USB to the virtual PC that is hosting the development software. The virtual PC encapsulates the experiment user interface. This is remotely accessed via RDP through a web browser.

To enable the remote PLC training described in [13], the GOLDi hardware architecture we extended as highlighted in Fig. 5:

- In addition to the GOLDi lab server software, the lab server machine runs now as virtual machine (VM) host software. Each virtual machine corresponds to a PLC device connected to the internal Ethernet network. The configuration of the VM host software guaranties that each start of an experiment causes resetting the VM to an initial configuration.

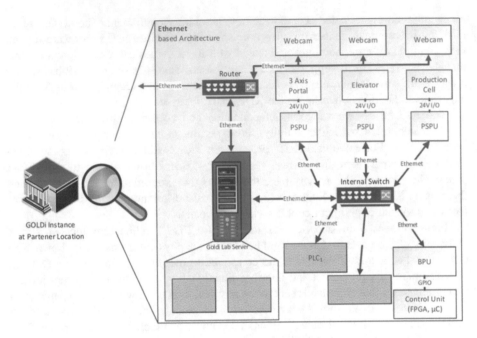

Fig. 5. GOLDi-Lab with external PLC device

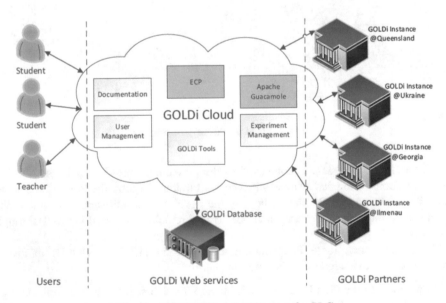

Fig. 6. GOLDi Cloud Architecture for PLC

- The configuration of the firewall allows the lab server VM only to communicate with its assigned PLC device.
- Each virtual machine has the manufacturer specific development environment pre-installed and is pre-configured for the use within the GOLDi Lab.

An experiment, using the PLC as a control unit, starts in the same manner as any other experiment within the GOLDi Lab. When configuring the experiment the student selects the PLC as a control unit. When the experiment is started, in addition to the ECP, another browser tab is opened in which the remote desktop of the virtual machine is displayed using Apache Guacamole. As displayed in Fig. 6, the user's machine is communicating with the GOLDi cloud via HTTP. The Guacamole instance within the GOLDi cloud then translates this communication to the RDP (Remote Desktop Proto-col) used by the virtual machines for remote desktop access.

Currently, work is underway to implement another GOLDi instance at the University of Southern Queensland in Australia to be able to establish additional PLC device based experiments.

6   Conclusion

To support the Smart Learning approach, we have analyzed in this paper the GOLDi-lab and found out how we can develop the next lab generation as a smart lab. Therefore, we first discussed the state of the art in developing smart labs and pointed out relevant properties. These properties we compared with the actual status of the GOLDi-lab and its architecture. Based on that we discussed a possible modularization of the lab to define and provide separate parts of the lab as services. Finally we gave an example of an ongoing project were we allow to use parts of the GOLDi-lab to create a new remote lab, using a PLC as control unit in connection with the GOLDi-lab COs and cloud services.

References

1. Uskov, V., Pandey, A., Bakken, J.P., Margapuri, V.S.: Smart engineering education, pp. 476–481, Piscataway, NJ, IEEE (2016)
2. Fäth, T., Henke, S.F., Henke, K., Wuttke, HD., Hutschenreuter, R.: On Effective Maintenance of Distributed Remote Laboratories. In: Proceedings of 2018 15th REV, Springer, Cham (2018)
3. Tawfik, M., Salzmann, C., Gillet, D., Lowe, D., Saliah-Hassane, H., Sancristobal, E., Castro, M.: Laboratory as a Service (LaaS). In 11th International Conference on Remote Engineering and Virtual Instrumentation (REV), pp. 11–20, Piscataway, NJ, IEEE (2014)
4. Lei, C.-U., Liang, H.-N., Man, K.L.: Building a smart laboratory environment at a university via a cyber-physical system, pp. 243–247, Piscataway, NJ, IEEE (2013)
5. Thompson, C.W.: Smart devices and soft controllers. IEEE Internet Comput. 9(1), 82–85 (2005)
6. Salzmann, C., Gillet, D.: From online experiments to smart devices, 4 (2008)

7. Salzmann, C., Govaerts, S., Halimi, W., Gillet, D.: The smart device specification for remote labs. Int. J. Online Eng. (iJOE) **11**(4), 20 (2015)
8. Henke, K., Fäth, T., Hutschenreuter, R., Wuttke, H.-D.: GIFT - an integrated development and training system for finite state machine based approaches. In: Online engineering & Internet of Things, vol. 22, pp. 743–757 .Springer, Cham (2018)
9. Knüpper, J., Seeber, M.: Browser based Digital Circuit Simulation, Ilmenau: TU Ilmenau (2017)
10. Henke, K., Ostendorff, S., Wuttke, H.-D.: A concept for a flexible and scalable infrastructure for remote laboratories. In: The Impact of Virtual, Remote, and Real Logistics Labs, vol. 282, pp. 13–24. Springer, Heidelberg (2012)
11. Henke, K., Ostendorff, S.: A grid concept for reliable, flexible and robust remote engineering laboratories. Int. J. Online Eng. iJOE **8**, 42–49 (2012)
12. Maiti, A., Zutin, D.G., Wuttke, H.-D., Henke, K., Maxwell, A.D., Kist, A.A.: A framework for analyzing and evaluating architectures and control strategies in distributed remote laboratories. IEEE Transactions on Learning Technologies, pp. 1–16 (2018)
13. Kist, A.A., Maiti, A., Maxwell, A.D., Henke, K., Wuttke, H.D., Fäth, T.: Enabling remote PLC training using hardware models. In: Proceedings of 2018 15th REV. Springer, Cham (2018)

PerspectivesX: A Collaborative Multi-perspective Elaboration Learning Tool

Aneesha Bakharia[✉] and Marco Lindley

Institute for Learning and Teaching Innovation, The University of Queensland,
Brisbane, QLD, Australia
aneesha.bakharia@gmail.com

Abstract. There currently exists a lack of scaffolded collaborative activities within MOOC's. In this paper, we introduce the PerspectivesX tool which has been designed to scaffold multi-perspective collaborative learning activities within MOOCs. The PerspectivesX tool is able to promote learner knowledge construction and curation for a range of multi-perspective elaboration techniques (e.g., SWOT analysis, Six Thinking Hats and activities with custom dimensions). The tool serves as a searchable knowledge base, is able to persist collective intelligence across course re-runs and clusters similar student responses together. In this paper, we illustrate that a good theoretical background already exists to support the design of CSCL tools and that via the LTI specification, feature-rich collaborative tools can be embedded within multiple courses. An evaluation of MOOC platform extension frameworks is presented to substantiate the implementation of the tool using the LTI standard. Key design principles, tool features, and proposed embedded analytics are also discussed.

Keywords: Computer supported collaborative learning
Massive open online courses · edX XBlock
Learning tools interoperability · Knowledge construction
Critical thinking · Idea generation

1 Introduction

The authoring tools within MOOC platforms promote the delivery of xMOOCs [12] by predominantly supporting the addition of text, video, quizzes and basic social polls (e.g., word clouds and multi-option polls). The wiki and discussion forum are usually the only collaborative tools. There currently exists a wide gap between the unstructured collaborative nature of forums and other MOOC instructional content (i.e., videos, quizzes and social polls). Within this paper, we introduce the PerspectivesX tool which has specifically been designed to address this gap and provide a way to scaffold learner collaboration.

© Springer International Publishing AG, part of Springer Nature 2019
V. L. Uskov et al. (Eds.): KES SEEL-18 2018, SIST 99, pp. 27–36, 2019.
https://doi.org/10.1007/978-3-319-92363-5_3

Recent research findings indicate that learners that actively contribute to the course forum, are more likely to complete the course and achieve higher grades [4]. However, only a small percentage of learners actively participate in a course discussion forum, with recent estimates of forum participation being between 5–10% of participants [8]. A large number of learners passively engage with a forum by reading posts and a larger percentage choose not to engage in the forum in any form. Tools that are able to scaffold collaborative learning activities by ensuring that learners share and review other learners ideas are therefore required.

An intriguing question is "Why is there a lack of scaffolded collaborative learning activities?". Plausible answers might be because a lack of collaborative learning activity theory exists or that the technology to support the implementation at either a small or large scale does not exist. Within this paper, we will illustrate that both the theory (via Knowledge Community and Inquiry model and CSCL) and technology (i.e., XBlocks and LTI) currently exist to support a range of scaffolded collaborative learning activities.

The PerspectivesX tool is able to scaffold a range of multi-perspective elaboration activities. The tool is designed to promote active participation from learners that are either not participating in a discussion forum or that are passive forum participants (i.e., only reading forum posts). PerspectivesX encourages learners to contribute and share ideas. Sharing is, however optional as learners can also choose to remain anonymous or not share their contributions with other learners. The knowledge base structure allows learners to explore, review and curate other learners submissions.

In a PerspectivesX activity, learners must think about a problem from an assigned or selected perspective and actively contribute their ideas to a knowledge base that is available to all course participants. Instructors can enable an optional curation layer that requires learners to collate ideas from fellow learners in order to complete the remaining perspectives of the activity. Curation is a 21st century digital literacy and has strategically been included in the tool. Curation has been shown to facilitate the development of learner search and evaluation strategies as well as promote critical thinking, problem solving, and participation in networked conversations [11].

A range of idea generation and multi-perspective elaboration activities can be created with the PerspectivesX tool including Strengths, Weakness, Opportunities and Threats (SWOT) analysis, Six Thinking Hats [5], Fishbowl [9] and SCAMPER [6]. The tool is also modular and flexible allowing instructors to define custom grid dimensions (i.e., perspectives).

The learning design objectives for the PerspectivesX tool include the ability to:

- Encourage students to submit and curate ideas across all perspectives
- Encourage students to start sharing ideas (even if they initially submit ideas as anonymous or not shared)
- Encourage students to curate a list of diverse ideas within and across perspectives
- Trigger discussion among learners in a post activity forum

PerspectivesX supports the following domain specific objectives:

- Encourage students to submit and curate ideas across all perspectives
- Encourage students to compile a comprehensive set of ideas (either submitted or curated) across all perspectives and submit original and innovative ideas.

2   Theoretical Background

The PerspectivesX tool implements concepts from CSCL scripting; and the Knowledge Community and Inquiry model (KCI) [13]. KCI uses Web 2.0 tools to add a layer of collective knowledge building to scripted learning activities. PerspectivesX implements the 4 main principles of KCI through its knowledge base of student perspective submissions (Principle 1), the inclusion of curation mechanics (Principle 2 & 3) and facilitation of instructor moderation (Principle 4) [13].

3   Design Guidelines

The design guidelines that have underpinned the development of the PerspectivesX tool are detailed below:

- **Support the design of structured knowledge construction, critical thinking and multi-perspective elaboration activities**
  PerspectivesX supports the collaborative construction multi-perspective elaboration grids. The grids are defined by an instructor and can include any number of dimensions (i.e., perspectives). Learners are required to contribute ideas and curate a diverse list of ideas across grid dimensions.
- **Provide the ability for students to receive a participation score**
  Discussion forum activity is often not graded within MOOCs making forum participation an opt-in activity for many students. In PerspectivesX, the instructor is able to assign grade percentages based on learner submission and curation.
- **Support opt-in and anonymous learner knowledge sharing**
  Learners are able to submit ideas and obtain a participation grade without being required to share their submissions. Sharing is an opt-in feature of the tool, enabled by default but easily able to be turned off by learners on individual submission items. Learners are also able to share their ideas anonymously. This is an important design guideline as learners may not be comfortable sharing their contributions. The aim of the activity is to encourage learners that are not sharing or sharing anonymously to start sharing their content and eventually participate in flow on discussion activities. If the learner decides not share their contribution, the contribution will still be included anonymously in summary word clouds and topic models.
- **Support instructor moderation**
  Moderators are also able to highlight interesting learner posts and correct misconceptions that are being curated. Moderator curated lists across dimensions help learners to focus their attention on relevant and diverse submissions [3] from other learners.

- **Support learner curation**
  Learners are able to view submissions shared by other learners within all dimensions (e.g., Strengths) and can curate (i.e., add an item to their own list) at any time. Learners are also able to view the number of times their submissions have been curated.
- **Support temporal independence**
  PerspectivesX is an asynchronous learning activity and is able to be embedded in both paced and self-paced MOOCs. Learners can submit and curate responses from other learners at any time. PerspectivesX can also be used in face-to-face lectures, workshops and tutorials to collate student ideas.
- **Support knowledge base growth and collective knowledge sharing across course re-runs and cohorts**
  Usually, activities are reset at the beginning of each course re-run with the new cohort of students starting from scratch. Valuable prior student discussions and ideas are lost. PerspectivesX allows the knowledge base of student submissions to be persisted across course re-runs, allowing students to build on the ideas proposed by the previous cohort. Retaining student contributions facilitates knowledge growth but also poses information retrieval problems. The interface used to display learner contributions, therefore, includes intuitive navigation and free text search. Similar student submissions are also clustered together, with the most definitive student submissions shown to prevent learners from reading repeated similar submissions from other students.
- **Facilitate the delivery of customized scalable feedback**
  PerspectivesX seeks to provide moderators with a high-level overview of student contributions within a dimension. Topic modeling algorithms are used to find the common topics within student submissions. A simple mechanism is provided for moderators to provide feedback to students based on both the topics they have and have not included. Best practice from other educational applications that use clustering techniques to provide feedback at scale [1,10] is also being included. Within PerspectivesX both the Latent Dirichlet Allocation (LDA) [2] and Non-negative Matrix Factorization (NMF) [14] algorithms are available. A summary of topics is included in the analytics dashboard.

4   Implementation Considerations: LTI Tool vs EdX XBlock

In this section, implementation technologies are evaluated. The PerspectivesX tool can either be implemented using the Learning Tools Interoperability (LTI) specification or as an XBlock for the Open edX platform. LTI tools can be built in any programming language, have their own user interface and are able to run on their own server. LTI tools are able to integrate with a range of Learning Management Systems that implement the LTI specification (i.e., Blackboard, Canvas, D2L, Moodle and Coursera). XBlocks are extensions for the Open edX platform, must be built in the Python programming language are adhere to the Open edX user interface standards. In Table 1, LTI is compared to the XBlock extension architecture in relation to the design requirements for the PerspectivesX tool.

Table 1. A comparison of implementing the PerspectiveX tool either as an LTI or XBlock

Feature	LTI	XBlock
Admin UI	A flexible user interface can be implemented in an LTI using any client-side framework	The user interface must adhere to the Open edX design guidelines. The admin interface will open in a modal window. Multi-step screens are cumbersome
Learner UI	A flexible user interface can be implemented in an LTI using any client-side framework. The learner user interface will be displayed within an iframe or open in a new browser tab	A flexible learner user interface can be built using client-side frameworks. The learner user interface will be embedded natively within the edX course page. The user interface will need to adhere to the Open edX design guidelines
Learner, instructor and moderator roles	The user role is passed to the LTI tool	Within an XBlock, the user role can be determined
Data persistence across course re-runs	The LTI context id will contain the course id for the current course. A custom parameter must be passed to the LTI that defines an id that is assigned to all instances of the course re-run	The XBlock architecture allows data to be persisted in various scopes. A course scope that persists across re-runs is available but this scope might not be flexible enough to store learner data across course re-runs
Ability to scale	The LTI tool must be designed to scale. The simplest way to scale would be to deploy the tool using Amazon Elastic Beanstalk	The Open edX platform is designed to be inherently scalable, provided that additional servers are added
Integration of NLP, deep learning and topic modeling algorithms	The LTI tool can be built in any programming language and integrate with any API. Most NLP and deep learning algorithms are available in the Python programming language, so it makes sense to program the tool using a Python web framework such as Django. NLP and deep learning algorithms are however compute intensive and would need to be executed on additional servers	XBlocks are programmed in the Python programming language. The advantage is that most NLP and deep learning algorithms are available in Python. As NLP and deep learning algorithms are compute intensive it is preferable that the algorithms are executed on additional servers and accessed via an API
Search and content indexing	Allowing learners to search is a feature that is required for courses with a large number of participants. Periodic indexing of student submitted content can be implemented in an LTI	Further investigation is required to determine whether student submitted contributions can be indexed on Open edX. Open edX does include forums that are indexed using Solr but this functionality may be difficult to implement within an XBlock

Both the LTI and XBlock implementation options are comparable except for the inclusion of advanced algorithms and search indexing. Implementing PerspectivesX as an LTI is the preferred option as it provides additional flexibility in the storage of student submissions and the ability to more readily integrate with NLP and topic modeling algorithms. Additionally, LTI tools are able to integrate with a variety of LMSs.

5   Prototype Implementation

PerspectivesX is open source and is programmed using Python, the Django web application framework and React. While most CSCL scripting tools use a flowchart metaphor [7], for simplicity and ease of use, PerspectivesX takes a declarative approach by providing the instructor with configurable options. The activity creation interface is shown in Fig. 1. The instructor is able to choose or create a new grid template and specify how learners contribute to the perspectives (i.e., dimensions) in an activity (i.e., the learner contributions section). Options are provided to allow learners to choose a perspective, contribute to all perspectives, or be randomly assigned to a perspective. Various curation options are also available, the instructor is able to decide whether a learners chooses the perspective or is assigned a perspective to curate.

Central to the design and flexibility of the PerspectivesX tool, is a structured template that instructors are able to create. The tool includes standard templates for common activities such as Six Thinking Hats [5], SCAMPER [6] and Fish Bowl [9]. Instructors are also to create custom templates. As an example, a template can be created for a SWOT activity using a multi-perspective fieldset to include each dimension that is required (i.e., Strengths, Weaknesses, Opportunities and Threats).

An example learner submission user interface is displayed in Fig. 2. In Fig. 2, the template is displayed as a grid and the learner can contribute ideas and select a sharing option (i.e., Shared, Anonymous and Not Shared) for each submission within a dimension. The learner is also able to view all learner submissions and able to add items submitted by other learners to their own list (i.e., curate items). The learner user interface clearly differentiates the content that the learner has submitted from the content they have curated. Learners are also able to export the grid they have created to a Microsoft Word document.

6   Embedded Analytics

An analytics dashboard for instructors and moderators is currently under development (see Fig. 3). The PerspectivesX tool provides an ideal opportunity for investigating analytics embedded directly within a learning tool and how analytics can be better mapped to learning design objectives.

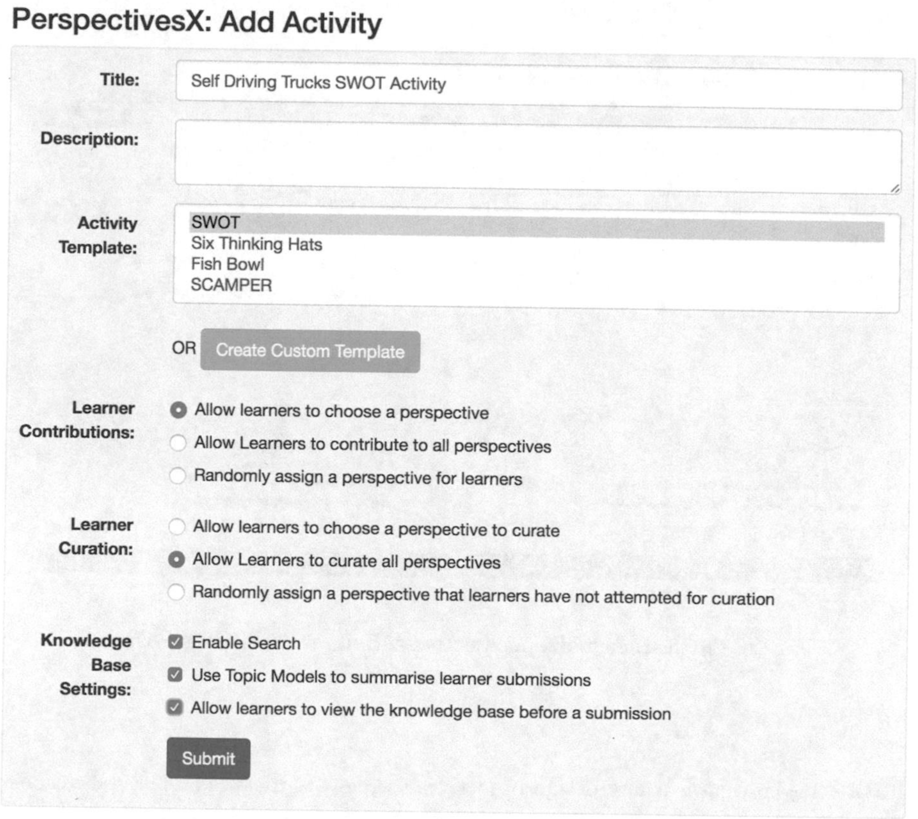

PerspectivesX: Add Activity

Title:	Self Driving Trucks SWOT Activity
Description:	
Activity Template:	SWOT Six Thinking Hats Fish Bowl SCAMPER
	OR Create Custom Template
Learner Contributions:	● Allow learners to choose a perspective ○ Allow Learners to contribute to all perspectives ○ Randomly assign a perspective for learners
Learner Curation:	○ Allow learners to choose a perspective to curate ● Allow Learners to curate all perspectives ○ Randomly assign a perspective that learners have not attempted for curation
Knowledge Base Settings:	☑ Enable Search ☑ Use Topic Models to summarise learner submissions ☑ Allow learners to view the knowledge base before a submission
	Submit

Fig. 1. The instructor multi-perspective activity creation interface.

While a variety of metrics and visualizations can be created with the data stored by the tool (i.e., the text of the idea, date of creation, number of times curated, learner details and the dimension/perspective the idea has been associated with), the learning design and domain specific objectives have been used to refine and focus on analytics that provide insight on whether the tool learning design objectives are being met. In Table 2, the learning design and domain specific objectives have been matched to appropriate metrics and visualizations.

A work-in-progress prototype of the analytics dashboard is shown in Fig. 3. A timeline showing the sharing options used by learners is displayed to help the instructor and moderators determine whether learner sharing behavior is changing as the activity is progressing. A distribution of posts is also shown for each dimension.

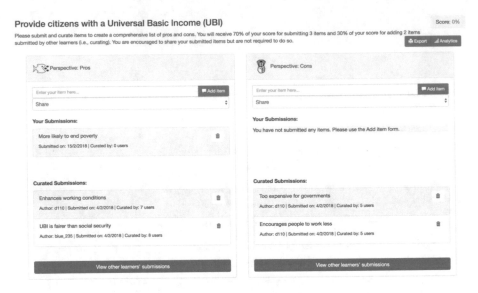

Fig. 2. The learner multi-perspective activity submission interface.

Table 2. Mapping learning design objectives to relevant metrics and visualizations.

Objectives	Metrics and visualizations to support objectives
Learning design objectives	
Encourage students to submit and curate ideas across all perspectives	Histogram of number of perspectives submitted to by students
Encourage students to start sharing ideas (even if they initially submit ideas as anonymous or not shared)	Timeline of submitted ideas by sharing options (i.e., see how sharing patterns change during tool use)
Encourage students to curate a list of diverse ideas within and across perspectives	Show topics covered by items submitted to a perspective and list of top curated items
Trigger discussion among learners in post activity forum	Sharing and Curation networks that lead to discussions
Domain specific objectives	
Students compile a comprehensive set of ideas (either submitted or curated) across all perspectives and submit original and innovative ideas	Show topics covered by items submitted to a perspective

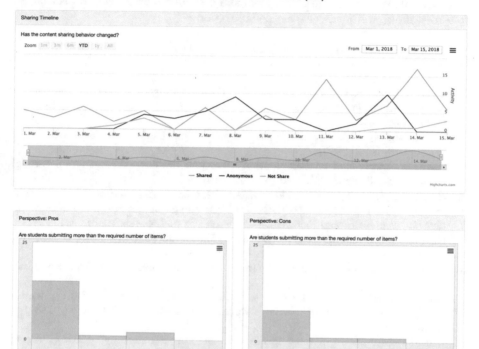

Fig. 3. The PerspectivesX instructor dashboard.

7   Conclusion and Future Directions

Within this paper, we were able to successfully illustrate that neither theoretical
or technical barriers are preventing the design and implementation of scaffolded
collaborative learning activities. PerspectivesX has been implemented to bridge
the gap between xMOOCs and cMOOCs [12]. The PerspectivesX tool includes
the ability to support structured knowledge construction, critical thinking and
idea generation activities; opt-in and anonymous learner knowledge sharing;
instructor moderation; learner curation, temporal independence (i.e., support
for self-paced MOOCs), knowledge base growth across course re-runs and is
able to take advantage of recent NLP techniques to provide customized scalable
feedback for learners.

Future research will focus on the evaluation of the PerspectivesX tool within
paced, self-paced and face-to-face on-campus activities. A key research question
being investigated is whether simple scaffolded collaborative activities are able
to help learners transition to become active participants in post activity forums.
A range of CSCL design patterns are also able to be implemented via the LTI

specification. We encourage researchers and developers to develop tools that are able to scaffold and foster learner collaboration.

References

1. Basu, S., Jacobs, C., Vanderwende, L.: Powergrading: a clustering approach to amplify human effort for short answer grading. Trans. Assoc. Comput. Linguist. **1**, 391–402 (2013)
2. Blei, D.M., Ng, A.Y., Jordan, M.I.: Latent Dirichlet allocation. J. Mach. Learn. Res. **3**, 993–1022 (2003)
3. Boyd, D.: Streams of content, limited attention: the flow of information through social media. Educause Rev. **45**(5), 26 (2010)
4. Corrin, L., de Barba, P.G., Bakharia, A.: Using learning analytics to explore help-seeking learner profiles in MOOCs. In: Proceedings of the Seventh International Learning Analytics and Knowledge Conference, pp. 424–428. ACM (2017)
5. De Bono, E., Pandolfo, M.: Six Thinking Hats, vol. 192. Back Bay Books, New York (1999)
6. Eberle, B.: Scamper On: Games for Imagination Development. Prufrock Press Inc., Waco (1996)
7. Faucon, L., Håklev, S., Hadzilacos, T., Dillenbourg, P.: Demo of orchestration graph engine: enabling rich social pedagogical scenarios in MOOCs, pp. 143–144. ACM (2017)
8. Hill, P.: MOOC discussion forums: barrier to engagement? (2013). http://mfeldstein.com/mooc-discussion-forums-barriers-engagement/
9. Miller, R.L., Benz, J.J.: Techniques for encouraging peer collaboration: online threaded discussion or fishbowl interaction. J. Instr. Psychol. **35**(1), 87–94 (2008)
10. Mohler, M., Mihalcea, R.: Text-to-text semantic similarity for automatic short answer grading. In: Proceedings of the 12th Conference of the European Chapter of the Association for Computational Linguistics, pp. 567–575. Association for Computational Linguistics (2009)
11. O'Connell, J.: Content curation in libraries: is it the new black? Collected Mag. **6**, 4–6 (2012)
12. Ping, W.: The latest development and application of massive open online course: from cMOOC to xMOOC. Mod. Dist. Educ. Res. **3**(005), 13–19 (2013)
13. Slotta, J.D., Najafi, H.: Supporting collaborative knowledge construction with Web 2.0 technologies, pp. 93–112. Springer, New York (2013)
14. Xu, W., Liu, X., Gong, Y.: Document clustering based on non-negative matrix factorization. In: Proceedings of the 26th Annual International ACM SIGIR Conference on Research and Development in Information Retrieval, pp. 267–273. ACM (2003)

Developing an Adaptive Mobile Platform in Family Medicine Field Experiences: User Perceptions

Christian Rogers[1(✉)], Corinne Renguette[1], Shannon Cooper[2],
Scott Renshaw[2], Mary Theresa Seig[3], and Jerry Schnepp[4]

[1] Indiana University-Purdue University Indianapolis,
Indianapolis, IN 46202, USA
`rogerscb@iupui.edu`
[2] School of Medicine, Indiana University, Indianapolis, IN 46202, USA
[3] Ball State University, Muncie, IN, USA
[4] Bowling Green State University, Bowling Green, OH, USA

Abstract. EASEL (education through application-supported experiential learning) is a platform designed to provide just-in-time content and reflection opportunities to students during field experiences, such as interviews or field labs, conducted as part of the workload in a course. This study was conducted in area of family medicine education at Indiana University-Purdue University Indianapolis. EASEL allows instructors and students flexibility to engage with course content based on the time of day and the location of each student conducting field work by providing access to questions and content before, during, and after a targeted field experience. In this study, three cohorts of family medicine students (N = 20) interviewed either a health care professional or a patient. Students used EASEL to facilitate and support their experience in the field. This study examined the student perceptions of EASEL. The data indicated instructive information on the usability of the EASEL platform and aided developers in considering future technologies to use as a part of the platform.

Keywords: Adaptive mobile learning experiential learning · Field experiences
Instructional design · User experience · Usability

1 Introduction

1.1 Supporting Just-in-Time Reflection

In the field of educational technology, a variety of tools exist to support experiential learning. Just as experiential learning experiences vary, so too do the tools used to support that - from learning management systems to individual apps. These tools allow instructors to adapt, support, and augment their instruction both in the classroom and outside the classroom. Students benefit as well inside the classroom, in the spaces of online learning, and in field experiences. Experiential learning allows them to ask questions, solve problems, and apply knowledge and abstract understanding to various learning environments such as the classroom, interviews, procedures, simulations, and

© Springer International Publishing AG, part of Springer Nature 2019
V. L. Uskov et al. (Eds.): KES SEEL-18 2018, SIST 99, pp. 37–50, 2019.
https://doi.org/10.1007/978-3-319-92363-5_4

experiments [1]. In this study, family medicine students participated in a field experience that requires them to prepare for an interview with both a patient and a health care professional, conduct the interviews, and debrief and reflect from the interviews. Historically, the preparation, debriefing, and reflection all occur outside of the interview process when the student is working in the learning management system at a computer. However, with the use of EASEL, the student perceptions of shifting to just-in-time content delivery and reflection within the platform were examined.

1.2   A New Tool

EASEL (education through application-supported experiential learning) is designed to be used by both students and instructors to optimize the learning experience. This is done by using both a web-based portal (for the instructor) and a mobile app (for the student). The advantage of utilizing a mobile platform allows for access to phone-specific features such as GPS tracking, camera operation, audio recording, voice-to-text processing, persistent network connectivity, and time. Users interact with the content and reflection prompts based on date, time, activity, and location. EASEL can deliver reflection prompts and content at salient times or when users reach specific locations before, during, and after a learning experience. Instructors can initiate experiences using the web-based portal. Specific content items and tasks completion are assigned under each experience. A task might consist of content to review, reflection questions to complete, an activity to track, or a photo or video that needs to be taken. An instructor can set parameters for when a student can access a specific task, such as when the student arrives at a location or after a specific date and time. The interface is, therefore, *adaptive* based on the location of the student and time of the experience.

A student can open EASEL, select the course and experience, and review the content or assessment items associated with that experience. For example, a student may be asked to watch a video or complete a question before conducting an interview. A student may be asked to audio record an interview or track the time. After all items are completed, the instructor can review them in the web-based portal.

Reflection and content prompts can target specific milestones throughout a learning activity by being triggered via GPS location, time of day, or a combination of the two. This can also be set at an individual level by the student or set by the instructor for a class experience. Students receive notifications to remind them to complete items. This allows students to anticipate content to review and reflection opportunities that can be entered in a variety of modalities including text, photos, audio, and video. Future iterations will likely integrate wearable technologies such as smart watches and augmented reality headsets to facilitate a more ubiquitous experience.

2   Literature Review

Changes in medical education have had significant effects on reflection activities, and a number of methods have been utilized to ensure and improve student reflection [2, 17, 18]. As students spend less time in the classroom and more time gaining valuable practical experiences, a platform that assists with the development of reflective learning through guided, real-world experiences provides the opportunity to gain knowledge through action, reflection and self-monitoring, and understanding the situation and the self so actions can change the next time a similar situation is encountered [3, 4, 18].

2.1 Experiential Learning in Medical Education

The Experiential Learning Theory was originally developed by David Kolb [3]. Kolb's model expanded on the idea of learning through discovery and experiences. The four stages of the model include (1) concrete experience, (2) reflective observation, (3) abstract conceptualization, and (4) active experimentation. The overall goal of the stages listed is for learners to reflect on experiences so they can put theory into practice later.

In the *Theories in Medical Education* series, Yardley stressed the importance of implementing experiential learning theory in medical education [5]. For many years, medical education was considered 'on-the-job' training, but with changing times, the need for learning from prior experience affects how students approach new experiences [5].

Many medical schools across the country incorporate experiential learning into their curricula [2, 6]. In a systematic review of research literature on experiential learning in nursing and medical education, a number of studies were found that discussed areas of medicine in which experiential learning was implemented. In many of those studies, it was understood that experiential learning helps students to share knowledge with an emotional connection to the experience but studies are limited in the contextual influences that hinder the development of reflective learning [7].

2.2 Values of Prompted Reflection

The second stage of Kolb's Experiential Learning Theory is reflective observation where learners reflect on the initial experiences. Reflection occurs often and has been shown to be beneficial in medical education [8, 18, 20].

Several methods of reflection occur including writing assignments, face-to-face feedback (debriefing), and written feedback. Students who complete debriefing following an exercise have a better understanding of their learning experience. Just-in-time adaptive intervention (JITAI) platforms change depending on the needs of the context and needs of the learners [10]. JITAIs have emerged recently in mobile health applications to assist patients with changes in behavior [9, 10]. Facilitating just-in-time content reminders and reflection prompts could benefit students with the reflection process by helping them to internalize their learning.

2.3 Mobile Learning and ANS

Tablets, smartphones, and wearable devices can enable on-demand access to learning resources from any time and location and provide notifications and reminders that easily integrate into the lives of the learners. Learners can engage with new opportunities beyond the traditional classroom and can participate more actively in their learning by being engaged in experiential and contextual learning that is embedded in real-life. Mobile learning can also offer real-time access to materials, communication, and exchange of knowledge with peers and experts in their field of study [11].

Adaptable navigation and personalized learning on mobile devices continues to increase because of the ease of mobility and flexible timing [12]. Mobile technology is being recognized as a way to facilitate learning and adapt to individual learner contexts. This has led to interest in the adaptation of content that can provide learning experiences that are tailored to the characteristics of the learner and the situations the learner is in. One example of this use is the *Units of Learning Mobile Player* [11], which supported partially automated adaptation of learning activities. The system helped the learner adapt to the activities (flow of learning) and adapted the educational resources, tools, and services for learning support systems. Gomez, Zervas and Sampson (2014) conducted a study utilizing the player and found that utilizing these methods of adaptation can facilitate student completion of the learning activities [11].

Most learning systems were created for computers and have been modified for mobile devices (rather than being created for mobile devices), which can restrict mobile functionality [13]. Learning management systems (LMS) offer mobile connections to content and communication and offer input such as games, simulations, and even audio/visual recording [14]. Within experiential learning, an LMS can be used to offer reflection opportunities through discussion boards, surveys, or the submission of document files. However, current LMSs fail to take advantage of mobile features that could aide in time-based or location-based prompts or allow students easy access to media use (camera/audio recording) during reflection opportunities. Taking advantage of these mobile features could eliminate the issue of delayed reflection entry and possibly improve student retention and retrieval of salient moments in the learning experience.

2.4   Just-in-Time Interventions

The idea of intervening at critical moments in the learning process is not new. Just-in-time teaching (JiTT) has been used throughout educational contexts and includes web-based, content-related questions for students to answer a few hours before class [15, 16]. It has been shown to increase student motivation and learning because instructors can tailor class content using the students' answers. JiTT was originally designed for face-to-face classroom use and has recently evolved to include additional components suitable for new environments able to advantage of JIT elements. One field that has adopted the idea of intervening just in time is that of mobile health and medicine. JITAI is a mobile health technological intervention that changes depending on the user's needs in a specific environment [18]. JITAIs have been shown to assist individuals with healthy behavioral changes including stress-reduction, smoking cessation, and increased activity [9, 10]. However, JITAI has not been widely used in education nor has it been studied in experiential learning contexts as a tool to foster opportunities for metacognitive thinking, and JITAIs are not currently focused on reflection tools. A platform that focuses on facilitating JITAI with relevant content and reflection prompts based on the time and location of the experience and allowing students easy access to mobile reflection tools could, therefore, assist students with retention and retrieval of salient moments in the learning experience. This could enhance opportunities for metacognitive development while reducing the load on working memory, leaving more resources for active learning.

3 Methodology

3.1 Context of the Study

Family medicine education introduces students to the principles and practice of caring for patients by allowing them opportunities to actively work with physicians in a community setting. This study was conducted within a family medicine clerkship at an urban institution. As a requirement of the clerkship experience, students must complete a Family & Community Project. The Family and Community Project serves as a capstone activity that has been designed to bring together what the medical students have learned during their clerkship while emphasizing the mission and values of Family Medicine, essentially applying theory to practice to culminate the experiential learning process. Students must identify a patient or population that faces barriers to health or healthcare, interview both the identified patient (or community resource representative) and a practicing health care professional who works with that patient or community and collaboratively find resources that might help the patient address one or more of those barriers. Students learn how to focus on discrete portions of a patient's medical history and physical concerns within the system of the patient's total health. Students in this clerkship setting meet with patients who present acute medical problems, are chronically ill, need preventive health education, or are seeking the support of their physician to cope with the trials and stresses of everyday life. Most importantly, students see patients interacting with their personal physician and witness the doctor-patient relationship as a learning experience. The goals of the Family and Community Project are for the students to

- Acquire a more comprehensive understanding about the impact of health and illness on a patient's life and family/support systems;
- Use a biopsychosocial approach to consider biological, psychological, and social factors and their complex interactions to better understand health, illness, and health care delivery to improve clinical patient care;
- Describe how the integration of community agencies, organizations, and other healthcare providers into patient care can positively impact health care outcomes;
- Develop strategies to use a team approach to develop a plan to improve health or health care outcomes.

For the assignment, students are asked to include the following elements:

1. The Patient's Story. The students will interview the patient to determine relevant background information and tell the patient's healthcare story in the assignment.
2. Barriers to Care. The students will describe at least one barrier to health or healthcare faced by the patient and his/her family. They will describe how the barrier impacts the patient's health and any strategies already in place to overcome the barrier(s).

3. Improving Health Outcomes. After meeting with the patient and/or his/her family, the students are to discuss how eliminating or mediating one or more of the barrier (s) can improve the patient's health or health outcomes. As such, students must collaborate with at least one other health professional or community resource person that interacts with or could interact with the patient or caregiver (i.e., pharmacist, social worker, nurse, teacher, daycare provider, dentist, etc.). This requires scheduling an appointment and an interview in order to work together with the healthcare professional.

4. Critical Thinking: By the end of the four-week clerkship experience, students will document and reflect upon their interaction with the patient and/or his/her family as well on the collaboration with the other healthcare professional.

Because of the nature and requirements of the clerkship, many students are faced with the challenges of meeting with the patient and healthcare professional at a time most convenient for those being interviewed. This often means squeezing the interview in between patients at the clinic or traveling off site. The information and tools for the students are currently housed as multiple PDF files or longer webpages within the course in a learning management system. Unfortunately, students have, at times, forgotten to download the PDF files prior to the interviews. Additionally, students have misplaced the papers after the interview due to the speed at which the clerkship runs. As well, the PDF documents are considered cumbersome to navigate by both students and instructors, but contain the information necessary for interpretation and interaction within the interview. Many students have not taken specific notes during a patient interview with the belief that they would remember the information later, only to see six new patients after the interview and not recall what exactly their interviewee said; as such, "which patient said that?" is a common thread.

At the completion of both interviews, the students are required to reflect on their experiences by completing a final project for the course. Students must take multiple pieces of information collected over three weeks and, then, during their fourth week, complete a comprehensive, reflective, capstone project. Due to the decentralized nature of the clerkship, the online learning management system (LMS) is heavily relied upon for storing course content and student assignments. The LMS also aims to increase the students' ability to become an active participant in his/her self-directed learning by keeping everything in a centralized course location. However, limitations of the LMS and its mobile app lead to some issues in supporting students as they participate in these important field experiences (for example, in scheduling their interviews, writing reflections, etc.), as students may not be able to quickly or easily get to a computer once an experience is complete.

The following study evaluated student perceptions of EASEL when used during the interview process. It was hypothesized that by utilizing the EASEL platform, family medicine students in this context would be able to quickly refer to information so they can provide relevant feedback during these interviews and would consider EASEL to be a helpful tool.

3.2   Procedures

The FMC is a required four-week clinical rotation for over 360 third-year medical students. The FMC is decentralized, assigning students to a family medicine physician in clinical locations throughout the state. Four days of each week on the rotation are spent with an assigned family medicine physician in his/her medical office. Each student with the FMC must complete two interviews. One interview is with a patient to understand the barriers to their healthcare. The other interview is with a healthcare professional to understand how to help overcome those barriers. After both interviews the student must complete a final project as a reflection piece based on the interviews.

Utilization of EASEL began in fall 2017. Three cohorts of students (N = 20) participated in the study. For one of the interviews students were encouraged (but not required) to review some questions in a document in the LMS before they conducted the interview. Those students who had an iPhone were asked to utilize EASEL for one of the interviews. When utilizing EASEL, the students followed the below procedure:

1. Students will schedule their interview in EASEL (see Fig. 1a) and review tasks (see Fig. 1c).

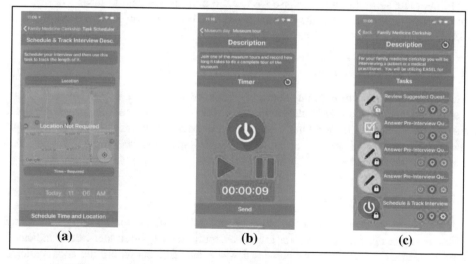

Fig. 1. (a) Scheduling Interview, (b) Tracking Time, (c) Task List

2. Students will receive a notification the day of their interview reminding them to complete two items: to review some questions about the interview and to answer a few pre-reflection questions.
3. Before the interview begins, students will use the time tracker to measure the time length of the interview (see Fig. 1b).
4. After the interview, students will use EASEL to respond to a few questions related to their interview experience (see Fig. 1c).

At the completion of each cohort, students who used EASEL (N = 20) were asked to complete a short survey including these two questions, which will be discussed in more detail in the *Data Analysis* section:

- What did you find most useful about EASEL? Please be specific.
- What would you suggest to improve EASEL?

Students who did not use EASEL (because EASEL is currently iOS only) were not included in the data results. The perceptions of using EASEL were evaluated to further inform future EASEL development.

4　Data Analysis

Students utilized EASEL for only one of their two interviews. For the interview not utilizing EASEL, students reviewed questions stored within a long document on the Canvas learning management system. They did not have a reflection opportunity for that interview. Future data analysis will include comparing the responses of those two surveys and looking at student final grades. Since course grades will not be completed until later this semester, and because we needed preliminary data to inform design and functionality improvements, the data analyzed here is focused on student perceptions of EASEL and areas that could be improved.

4.1　Most Useful Features

Students were asked to explain what they believed to be most useful about EASEL. Some students utilized words like *reminder*, *guide* or *motivator* to indicate what helped them with their experience. Other students noted specific features such as *the ability to track the time of an interview* or *to reflect on the interview*. See Table 1 for selected qualitative feedback.

4.2　Areas for Improvement

Students were asked to provide suggestions on how EASEL could be improved. Many of the responses indicated needs for improvement to the user interface. Others indicated a need to adjust the content and content delivery mechanisms within the app. Students also indicated issues with bugs that need to be mitigated and others indicated a general belief that the EASEL app was unnecessary.

Improvements to the User Interface. Students provided feedback on the user interface both in design and functionality. Improvements included accessing the app, structure, navigation of the app, input of content (such as question responses) and understanding the timer. See Table 2 for selected qualitative feedback.

Content. Students provided feedback on the content of the assignments as well as the delivery mechanisms for those assignments. Students indicated issues with questions (whether it be not understanding Likert scale questions, wanting to input their own questions or the inclusion of questions for the patient). One student also indicated a

Table 1. Student feedback indicating what was most useful about EASEL

Most useful item	N	Qualitative feedback
Reminder	5	"I think the EASEL app can be useful to remind us of when the meeting with the patient is." "I liked that the app asked what question I would ask first during the interview because it helped me prepare, but more questions like that would be even better."
Guide/Clarification	7	"It gave a chance for me to organize my thoughts" "It guided my questions and guided me through the process of interviewing." "It gave specific questions for me to ask and helped me formulate good rapport with my patient as it asked questions from her perspective."
Reflection/Other features	2	"I found the post-interview questions spaced out into their own boxes and questions the most helpful." "…timer for the interview, recording the responses of the interview"
Motivator	2	"Motivated me to read more about my patient prior to the interview and to take more time to think about my expectations for the interaction." "I really liked the sample questions which allowed me to think about things I have not thought about before. It broadened the spectrum of my interviews, I think."
Ease of use	2	"Easy to understand the app, Easy to find the resources" "The list of suggested questions was easy to access"

desire for an audio recorder and three other students stated there were issues with tasks being locked at certain times. See Table 3 for selected qualitative feedback.

Bugs/General User Experience Issues. Some students indicated issues with bugs in the app, while other students indicated that there was no need for EASEL. One student did not indicate a need for improvement of the EASEL. See Table 4 for data and selected qualitative feedback.

5 Discussion and Conclusion

Kolb's theory of experiential learning contains reflection, synthesis, decision making, accountability, and an experience that allows for natural trial and error. For years, experiential learning and reflection have been implemented in medical education, with many studies showing advances that could lead to better patient care. While a small sample size, this study showed that using an adaptive platform for just-in-time content delivery and reflections on experiences in medical education could be beneficial while also providing valuable feedback on an early version of the EASEL platform. While a

Table 2. Student feedback about EASEL's user interface

Areas of improvement	N	Qualitative feedback
Accessing the app		"I had difficulty with access to the application at the time of interview. I navigated the application interview afterwards. In retrospect, the application most likely would have helped as a reminder of the questions to ask in the interview."
Structure	1	"I think it needs to have more structure. I was having some difficulty using the app at first."
Navigation	8	"When I navigated the application after my interview, it was difficult to move on to the next section. Because I came prepared with the same tools the application offered, I am not sure the existing format enhances the experience of the interview." "It is also inconvenient that each question is on a separate "link" instead of just being able to answer all the pre-interview questions in one "link." I ended up not using the whole app because there were too many steps that were not needed to get the most out of this project."
Input	5	provide a space for jotting down thoughts/answers to each individual question rather than providing a single space at the end of a list of 10 + questions. Even though it wasn't required to answer each one, it just felt like an enormous amount of typing to be done on a cell phone to sufficiently complete that task." "Having a text box after each question vs. having just one large box at the bottom. It was harder to stay organized when bouncing from the top of the screen to the bottom." "I would increase the font size. Some of the questions required long responses and it's not user-friendly to type in such long answers on a phone."
Timer	1	"The timer was also confusing, since I could not see the seconds (clock displayed as 00:00…). I could not figure out when the timer was started or stopped (unless I were to stare at the screen for a full minute, which I later did in trying to get it to let me move on to the next question) and ended up not recording the time of my interview."

few participants acknowledged that EASEL was at least somewhat helpful, there was a strong indication to improve both the interface, delivery of content and the need to mitigate bugs.

5.1   Positive Features of EASEL

The most noted positive feedback from the study indicated that EASEL can be a helpful reminder tool during an interview process and that EASEL can serve as a guide during the process and can even motivate students to prepare more for their interview.

Table 3. Student feedback about content

Areas of improvement	N	Qualitative feedback
Questions	3	"The "Answer Pre-Interview Question -1" lists how prepared were you for the interview indicating 4 is the highest, but the options are in A-D with not numbers or text beside the options. Just an fyi, so I chose C, but am not sure what that indicated." "One recommendation to improve the application is to have the questions pertinent to the chief health are barrier accompanied by both quantitative and qualitative response sections so as to ease one's ability to take notes if one were to use the application as the interview unfolds. Another recommendations to improve the application is to possibly have a section in which the patient may provide feedback to one's performance, which may then subsequently be provided to us by the clerkship coordinator at a later date."
Audio recording	1	"Feature to record the interview"
Locking tasks	3	be useful to not lock them until viewed in order, so if other people experience this problem, they would still be able to utilize the other sections, rather than being stuck as I was."

Table 4. Student feedback about bugs and glitches with using EASEL

Areas of improvement	N	Qualitative feedback
Bug - Crashing	4	"The app crashed several times throughout the interview and I ended up not using the app to the full potential. I tried again via telephone call with the patient but again it crashed. I even put the location settings on correctly:"
General glitches	2	"theres [sic] glitches. Sometimes it would go back to a question I just answered and I did not want to type the entire response again. Also I don't think it's appropriate to be using your phone during an interview"
Unnecessary use	4	"I did not find it useful. Felt like transcribing the interview unnecessarily."
Unclear	1	"Easier to use when timing the interview. It was unclear when to use the app so I filled out all the information post interviews"
No feedback	1	N/A

One student indicated "I liked that the app asked what question I would ask first during the interview because it helped me prepare, but more questions like that would be even better." Responses such as this have been validated by early studies (Schnepp and Rogers [19]) which examined early perceptions of EASEL with a low fidelity prototype.

5.2 Areas for Improvement

The previous study presented data from the first native version of the EASEL app. Upon completion of the native iOS app, the team knew that there would be issues related to user interface, content delivery, and bugs or glitches. Students specifically noted issues with navigation where after completing a task they were not taken to the previous screen but to a home screen where they would then need to go through the process again. Other students indicated issues with typing large amounts of text onto a small screen. This feedback indicates a possible need for revisions in the interface for phones or potentially to allow students the use of a tablet during the process if content questions require long answers.

It was also noted that students felt they needed to use EASEL during the interview to take notes and they would have to be on their phone which was not the intention of EASEL. Content of assignments was a concern for some students as well, so the EASEL team plans to implement ideas for both instructors and students to help assignment creation, instructions, and completion to optimize EASEL's effectiveness.

Other students indicated issues with bugs such as crashes and general glitches. Four students stated they felt the EASEL app to be unnecessary. Some of this response may have been driven by the frustration with glitches. Others may have stated this due to their struggle with inputting text on a phone while others may have misunderstood the use of the app (i.e. using the app during the actual interview).

Some of the feedback from the students were a function of the environment EASEL was being utilized in. Students were conducting patient interviews and thus they would not have been allowed to record the audio of the interview for privacy reasons (even though an audio recorder is available). These suggestions will be fully utilized in the development of the next iteration of EASEL.

There is limited research related to using just-in-time adaptive reflection through mobile technology in medical education. Of the literature found, most focused on using mobile applications for direct behavior-changing applications. The literature review demonstrates a gap in using just-in-time adaptive reflection for experiential learning reflections in medical education. While this preliminary study offers a potential platform and data indicates a potential for EASEL to fill that gap, it also offers a glimpse of what users found to be useful when using EASEL and some areas of potential improvement for future iterations of the EASEL platform. The information gained from this study can be applied in current medical education through a number of avenues. One example is the use of EASEL to help support student engagement in field experiences with pre-experience, experience, and post-experience contexts. This study opens several areas for additional research. This interdisciplinary team of researchers and developers experienced their own learning through this process, and will continue collaborating on this innovative curricular development and technological intervention project.

References

1. Association for Experiential Education. http://www.aee.org/what-is-ee. Accessed 23 Dec 2016
2. Nutter, D., Whitcomb, M.: The AAMC Project on the Clinical Education of Medical Students. Association of American Medical Colleges, Washington, DC (2001)
3. Kolb, D.A.: Experiential Learning: Experience as the Source of Learning and Development. Prentice-Hall, Englewood Cliffs (1984)
4. Kolb, A.Y., Kolb, D.A.: The learning way: meta-cognitive aspects of experiential learning. Simul. Gaming **40**(3), 297–327 (2009)
5. Yardley, S., Teunissen, P.W., Dornan, T.: Experiential learning: transforming theory into practice. Med. Teach. **34**(2), 161–164 (2012). https://doi.org/10.3109/0142159X.2012.643264
6. Lin, S.Y., Schillinger, E., Irby, D.M.: Value-added medical education: engaging future doctors to transform health care delivery today. J. General Intern. Med. **30**(2), 150–151 (2015). https://doi.org/10.1007/s11606-014-3018-3
7. Allodola, V.F.: The effects of educational models based on experiential learning in Medical Education: an international literature review. Tutor **14**(1), 23–49 (2014). https://doi.org/10.14601/Tutor-14725
8. Lewin, L.O., Robert, N.J., Raczek, J., Carraccio, C., Hicks, P.J.: An online evidence based medicine exercise prompts reflection in third year medical students. BMC Med. Educ. **14** (164), 1–7 (2014). https://doi.org/10.1186/1472-6920-14-164
9. Nahum-Shani, I., Hekler, E.B., Spruijt-Metz, D.: Building health behavior models to guide the development of just-in-time adaptive interventions: a pragmatic framework. Health Psychol. **34**(Supplement), 1209–1219 (2015)
10. Nahum-Shani, I., Smith, S.N., Spring, B.J., Collins, L.M., Witkiewitz, K., Tewari, A., Murphy, S.A.: Just-in-time adaptive interventions (JITAIs) in mobile health: key components and design principles for ongoing health behavior support. Ann. Behav. Med. (2016). https://doi.org/10.1007/s12160-016-9830-8
11. Gómez, S., Zervas, P., Sampson, D.G., Fabregat, R.: Context-aware adaptive and personalized mobile learning delivery supported by UoLmP. J. King Saud Univ. Comput. Inf. Sci. **26**(1), 47–61 (2014)
12. Fakeeh, K.A.: Mobile-learning realization and its application in academia. Int. J. Comput. Appl. **133**(17), 8–16 (2016)
13. Subramanian, V., Rajkumar, R.: Investigation on adaptive context-aware m-learning system for teaching and learning basic Hindi language. Indian J. Sci. Technol. **9**(3), 1–8 (2016)
14. Leinonen, T., Keune, A., Veermans, M., Toikkanen, T.: Mobile apps for reflection in learning: a design research in K-12 education. Br. J. Edu. Technol. **47**(1), 184–202 (2016)
15. Novak, G.M.: Just-in-time teaching. New Dir. Teach. Learn. **2011**(128), 63–73 (2011)
16. Simkins, S., Maier, M.: Using Pedagogical Change to Improve Student Learning in the Economics Major, Chapters. In: Educating Economists, chapter 8 Edward Elgar Publishing (2009)
17. Ng, S., Kinsella, E., Friesen, F., Hodges, B.: Reclaiming a theoretical orientation to reflection in medical education research: a critical narrative. Med. Educ. **49**, 461–475 (2015)
18. Sandars, J.: The use of reflection in medical education: AMEE Guide No. 44. Med. Teach. **31**, 685–695 (2009)

19. Schnepp, J.C., Rogers, C.: Evaluating the Acceptability and Usability of EASEL: A Mobile Application that Supports Guided Reflection for Experiential Learning Activities (2017). Jite.org

20. Aronson, L., Niehaus, B., Hill-Sakurai, L., Lai, C., O'Sullivan, P.: A comparison of two methods of teaching reflective ability in Year 3 medical students. Med. Educ. **46**, 807–814 (2012). https://doi.org/10.1111/j.1365-2923.2012.04299.x

The Quality of Text-to-Voice and Voice-to-Text Software Systems for Smart Universities: Perceptions of College Students with Disabilities

Jeffrey P. Bakken[1,3](✉), Vladimir L. Uskov[2,3], Narmada Rayala[2,3],
Jitendra Syamala[2,3], Ashok Shah[2,3], Lavanya Aluri[2,3],
and Karnika Sharma[2,3]

[1] The Graduate School, Bradley University, Peoria, IL, USA
jbakken@fsmail.bradley.edu
[2] Department of Computer Science and Information Systems, Bradley
University, Peoria, IL, USA
uskov@fsmail.bradley.edu
[3] InterLabs Research Institute, Bradley University, Peoria, IL, USA

Abstract. Smart Universities and Smart Classrooms are the wave of the future. To better educate local and distant college students we will need to approach education and how we teach these students differently. In addition, college students are more technological than ever before and are demanding new and innovative ways to learn. This paper presents some ideas about how college students with disabilities might also benefit from Smart Classrooms and smart systems – especially from software systems. Even though students with disabilities are not the majority of learners in our classes, by incorporating university-wide smart systems and technologies we believe many of these students will also benefit. This paper presents the outcomes of a pilot research study analyzing two different commercially available and open source text-to-voice software systems and two different voice-to-text software systems by actual college students with disabilities. It describes (1) testing data obtained from actual college students with disabilities analyzing text-to-voice and voice-to-text software systems, (2) student suggestions for these types of systems for Smart Universities to consider, and (3) the impact these software systems could have on the learning of students with disabilities and how this software could aid universities to a possible transformation from a traditional university into a smart one.

Keywords: Smart classroom · Students with disabilities · Text-to-voice
Voice-to-text · Software systems

1 Introduction

Smart universities and smart classrooms can create multiple opportunities for students to learn material in a variety of ways. In addition, they can give student who would normally not have access to these materials opportunities to interact with the materials as well as the professor and other students. Although not designed or even conceptualized

© Springer International Publishing AG, part of Springer Nature 2019
V. L. Uskov et al. (Eds.): KES SEEL-18 2018, SIST 99, pp. 51–66, 2019.
https://doi.org/10.1007/978-3-319-92363-5_5

to benefit students with disabilities, this concept would definitely have an impact on the learning and access to material for students with all different types of disabilities.

2 Literature Review

2.1 Smart Universities: Literature Review

Primary focus of smart universities is in the education area, but they also drive the change in other aspects such as management, safety, and environmental protection. The availability of newer and newer technology reflects on how the relevant processes should be performed in the current fast changing digital era. This leads to the adoption of a variety of smart solutions in university environments to enhance the quality of life and to improve the performances of both teachers and students. Nevertheless, we argue that being smart is not enough for a modern university. In fact, all universities should become smarter in order to optimize learning. By "smarter university" we mean a place where knowledge is shared between employees, teachers, students, and all stakeholders in a seamless way [1].

A smart campus ranges from a smart classroom, which benefits the teaching process within a classroom, to an intelligent campus that provides lots of proactive services in a campus-wide environment". Although features, components, and systems of a smart university taxonomy have been discussed [2], only one publication could be located that discussed smart universities, smart classrooms and students with disabilities [3] and one publication could be located that discussed software systems for students with disabilities [4]. Given that 10% of all students have disabilities this is definitely an area that needs a more thorough investigation.

2.2   Speech-to-Text and Text-to-Speech Software Systems for Students with Disabilities: Literature Review

Only a handful of studies on college students and software systems could be located. They will be described next. Stinson et al. [5] conducted two investigations with 48 deaf and hard-of-hearing high school students and 48 deaf and hard-of-hearing college students. All participants viewed one lecture with an interpreter and one with the C-Print speech-to-text support service. High school students retained more lecture information when they viewed speech-to-text support as compared to interpreter support and when they studied note taker notes or a hard copy of the text after viewing the lecture. For college students, however, there was no difference between retention with these two kinds of support or with study of notes, compared to no study.

Ryba et al. [6] examined continuous automated speech recognition in the university lecture theatre. The participants were both native speakers of English (L1) and English as a second language students (L2) enrolled in an information systems course (Total N = 160). After an initial training period, an L2 lecturer in information systems delivered three 2-h lectures over a three-week period to the participants and other students. Student self-reports indicated that there were a number of perceived benefits associated with the use of continuous automated speech recognition. Compared with

L1 students, a significantly greater number of L2 students and students with special needs reported that the system had potential as an instructional support mechanism. However, a greater accuracy in the system's recognition of lecture text vocabulary needs to be achieved. The implications are that lecturers need an extensive training period before delivering lectures using continuous automated speech recognition.

Nelson and Reynolds [7] examined the composing processes of five postsecondary students with disabilities who were learning to use speech recognition software for college-level writing. The study analyzed their composing processes through observation, interviews, and analysis of written products. Results indicated that speech recognition software was an effective writing technology for students with disabilities. In addition, some students who used this speech recognition software had a strong aversion to formal planning and the authors believe this software is not yet suitable for general use in a college composition classroom.

Glasser et al. [8] described the accessibility challenges in using the top seven most popular Automatic Speech Recognition (ASR) applications on personal devices for commands and group conversation, by five participants, aged 19 to 47 years old, who either had hearing or were deaf or hard of hearing, including the authors. The authors presented the most common use cases, their challenges, and best practices plus pitfalls to avoid in using personal devices with ASR for commands or conversation.

Iglesias et al. [9] conducted a Spanish educational project that aimed at improving inclusive education for all. It proposed two main accessible initiatives: (1) real-time captioning and text-to-speech (TTS) services in the classroom, and (2) accessible Web-learning platforms out of the classroom with accessible digital resources. The authors provide an evaluation of the into-the-classroom initiative (real-time captioning and TTS services). This evaluation has been conducted during a regular undergraduate course at a university and during a seminar at an integrated school for deaf children. Forty-five hearing students, 1 foreign student, 3 experts in captioning, usability and accessibility, and 20 students with hearing impairments evaluated these services in the classroom. Evaluation results show that these initiatives are adequate to be used in the classroom and that students are satisfied with them.

Izzo et al. [10] examined the effects of a text-to-speech screen reader program on the academic achievement of high school students with disabilities in an online transition curriculum emphasizing information literacy. The text-to-speech support was introduced and withdrawn in a reversal design across 10 curriculum units. Findings suggest that the text-to-speech support increased unit quiz and reading comprehension performance with large effect sizes.

MacArthur and Cavalier [11] addressed the feasibility and validity of dictation using speech recognition software and dictation to a scribe as accommodations for tests involving extended writing. On the issue of feasibility, high school students with and without learning disabilities (LD) learned to use speech recognition software with acceptable accuracy. Total word errors with speech recognition were under 10%, and there were few unreadable words. On the issue of writing quality, for students with LD, essays dictated using speech recognition were better than handwritten essays, and essays dictated to a scribe were even better. No differences in quality were found for students without LD. The results provide support for the validity of dictation as a test accommodation.

The outcomes of the analysis of these and several other related recent publications clearly show researchers in various schools and research center are trying to test and evaluate various Automatic Speech Recognition (ARS) and Text-to-Speech (TTS) systems by actual students with disabilities in lab or school environments. This analysis motivated us to go even further: our research project is focused on the benchmarking of ARS and TTS commercial and opens-source (free) systems by actual university students with various types of disabilities in a real campus-based learning environment on computers that are used in computer labs by all types of students.

2.3   Research Project Goal and Objectives

The performed analysis of above-mentioned and multiple additional publications and reports relevant to (1) smart universities, (2) university-wide smart software and hardware systems and technologies, (3) smart classrooms, (4) smart learning environments, (5) smart educational systems, and (6) students with disabilities undoubtedly shows that SmU-related topics will be in the focus of multiple research, design and development projects in the upcoming 5–10 years. It is expected that in the near future SmU concepts and hardware/software/technological solutions will start to play a significant role and be actively deployed and used by leading academic intuitions in the world.

Project Goal. The overall goal of this ongoing multi-aspect research project is to find out what actual college students with disabilities think in regards to the functions and appeal of open source as well as commercially produced software programs in the areas of text-to-voice and voice-to-text. This paper presents the outcomes of a pilot research study analyzing two different text-to-voice software systems and two different voice-to-text software systems (commercially available and open source) by actual college students with disabilities. This chapter will highlight: (1) results from actual college students with disabilities analyzing text-to-voice and voice-to-text software systems, (2) suggestions in these areas for Smart Universities to consider, and (3) the impact this software could have on the learning of students with disabilities and how this software could aid universities to a possible transformation from a traditional university into a smart one.

Project Objectives. The objectives of this project were to have actual college students with disabilities examine and evaluate open source and commercial software programs in the areas of (1) text-to-voice and (2) voice-to-text that could be implemented in a Smart University Classroom to aid students with disabilities (and possibly students without disabilities).

3   Students with Disabilities

Students in our classrooms may experience a variety of different types of disabilities. These possible disabilities include: learning disabilities, speech or language impairments, visual impairments, hearing impairments, attention deficit disorders, emotional or behavioral disorders, cognitive impairments, or physical disabilities. For example, at Bradley University there are approximately 410 students that have been diagnosed with these disabilities and have identified themselves so that they may receive services through the university. In addition, there are probably others that have not identified themselves and still others that may have not been diagnosed with disabilities. Given the difficulties that students with disabilities encounter during their lives and in school the software systems in Smart Classrooms and Smart Universities would benefit them and help them learn more efficiently and effectively and in many cases allow them to interact better with their professor and classmates. Where traditional classrooms do not specifically address software systems and how students with disabilities could be impacted, the implementation of specific software systems in smart classrooms and learning environments would definitely have an impact on the learning of these individuals and the difficulties students with disabilities encounter. The implementation of specific software systems could address the exact areas that are of difficulty for students with disabilities and allow them to fully participate in the classroom learning environment. A list of possible software systems that may benefit students with various types of disabilities and will be addressed later in this chapter are listed in Table 1.

Table 1. Students with disabilities and software systems that may be beneficial [3]

Types of students with disabilities	Software systems that may benefit students with disabilities
Learning disabilities	Text-to-Voice and Voice-to-Text software systems
Speech or language impairments	Text-to-Voice and Voice-to-Text software systems
Visual impairments	Text-to-Voice and Voice-to-Text software systems
Hearing impairments	Voice-to-Text software systems
Attention deficit disorder	Text-to-Voice and Voice-to-Text software systems
Emotional or behavioral disorders	Text-to-Voice and Voice-to-Text software systems
Cognitive impairment	Text-to-Voice and Voice-to-Text software systems
Physical disabilities	Text-to-Voice and Voice-to-Text software systems

4 Text-to-Voice Software Systems

There are many available text-to-voice software systems that could be implemented in a Smart Classroom within a Smart University. This software will allow the user to convert text to voice so they can hear what information the text is trying to convey if they have issues with reading and comprehending text. Instead of students focusing on reading the text they can focus on comprehending it. For example, the act of reading for some students is a cognitive process. These students see words and have to figure out what letters are in the words, what the letters sound like, and what the actual word is so all their energy is spent on the task of reading, not comprehending the material. Using this software will make the material more accessible to the student with these difficulties. For other students, the actual act of reading is automatic and they can focus on comprehending what they are reading. See Table 2 for a list of desired features for text-to-voice software.

Table 2. A list desired features of Text-to-Voice software systems for smart universities [3]

Desired system features	Feature details
1. Quick response	The system should convert text-to-voice instantly
2. Proof reading	Student or faculty should be able to listen to their notes or assignments, in order to improve the quality of information
3. Access on mobile-devices	It should allow users to convert text-to-voice anywhere
3. Drag-and-Drop	This option should allow users to drag their external files to the software, so that it reads aloud for them
4. Multi-linguistic	The software should support several popular languages
4. Highlight word	The word that is read aloud should be highlighted
5. Pronunciation editor	Manually modify the pronunciation of a certain word
5. Batch convertor	Convert multiple documents to mp3, wav, wma etc.
6. Type and talk	A mute student should be able to communicate easily by simply typing what he/she wants to say
7. High quality	Speech should be of high quality with clear pronunciation and minimal errors

After investigating desired and actual features of text-to-voice software systems that could be implemented in a Smart Classroom within a Smart University the next step was to evaluate the top three open source and commercial software programs. The programs chosen (based on our analysis) to implement with actual students with disabilities were Natural Reader (commercially available) and Windows Speech (Open Source).

5   Voice-to-Text Systems

There are many available voice-to-text software systems that could be implemented in a Smart Classroom within a Smart University. This software will allow the user to convert their voice to text if they have issues with written expression. Instead of students focusing on the actual writing process they can focus their attention on producing a high quality product. For example, the act of writing for some students is a cognitive process. These students think of a word, have to think of the letters that make up this word, and then have to think of how the letter looks so they can retrieve it from memory and write it down. This process is very time consuming and by the time they have written a few words they have lost their thoughts on what they initially had planned to write. Using this software will allow the student with a disability more access and the ability to produce higher quality written products. For other students, the actual act of writing is automatic (i.e., letter formation, word spellings, punctuation, etc.) and they can focus on the content of the message or assignment they are involved in writing. See Table 3 for a list of desired features for this software in existing Voice-to-Text software systems for Smart Classrooms and Smart Universities.

Table 3. A list desired features of Voice-to-Text systems for smart universities [3]

#	Desired system features	Details
1	Dictate continuously	Help faculty to dictate notes continuously in a normal, conversational pace without slowing down pace or over-enunciating words
2	Robust documentation	Should allow any user to create documents with punctuation marks
3	Accent support	Should allow faculty from different locations to communicate easily
4	Hands-free	It should help students with disabilities such as repetitive strain injury (RSI), dyslexia, vision impairment etc
5	Recognition speed	The text should appear on screen as it is dictated, without any delay
6	Accuracy	The text should be accurate without any major errors
7	Mobility	Documents should be easily integrated with cloud technology
8	Web search	Students should be able to search the web by just dictating
9	Multi-lingual	System should be able to listen to text in native language voices
10	Easy-to-use	Help users to dictate and when finished, can simply copy-paste dictated text where needed

After investigating desired and actual features of voice-to-text software systems that could be implemented in a Smart Classroom within a Smart University the next step was to evaluate the top three open source and commercial software programs. The programs chosen (based on our analysis) to implement with actual students with disabilities were Dragon Naturally Speaking (commercially available) and Google Docs (Open Source).

6 Research Environment and Questionnaire

For each software system, college undergraduate students with disabilities participated in two sessions. The first session, was an hour long training session. The second session, held at a different time (for the first two students and then combined for the remaining 5 students), was for two hours and consisted of one 30-min session working with each of the four systems: (1) Natural Reader, (2) Windows Speech, (3) Dragon Naturally Speaking, and (4) Google Docs. Each of the sessions was monitored and directed by two graduate students. After the conclusion of each 30-min session students were asked to evaluate each system using the categories listed in Table 4.

Table 4. Evaluation components for Text-to-Speech and Speech-to-Text software systems

Text-to-Voice software systems	Voice-to-Text software systems
Learning/Understanding the System	**Learning/Understanding the System**
• Learn system's main purpose	• Learn system's main purpose
• Learn system's main functions	• Learn system's main functions
• Understand Graphic User Interface (GUI) features	• Understand Graphic User Interface (GUI) features
• Understand HELP sub-system	• Understand HELP sub-system
Working with the System	**Working with the System**
• Main System's functions	• Main System's functions
• Systems' Graphic User Interface GUI features	• Systems' Graphic User Interface GUI features
• System's HELP sub-system	• System's HELP sub-system
User Opinions	**User Opinions**
• System useful for you?	• System useful for you?
• System advantageous for you?	• System advantageous for you?
• Overall rating of system?	• Overall rating of system?
• Preference of this system in daily life?	• Preference of this system in daily life?
	• Rating of speech recognition quality of this system?
Recommendations	**Recommendations**
• Any additional features recommended?	• Any additional features recommended?
• Any additional GUI features recommended?	• Any additional GUI features recommended?
• Any additional HELP sub-system features recommended?	• Any additional HELP sub-system features recommended?
• Any other recommendations to improve functionality of the system?	• Any other recommendations to improve functionality of the system?

7   Research Findings and Outcomes

Results of the evaluations completed by each of the participants are described in Tables 5 and 6. Text-to-Voice software systems were Natural Reader (commercially available) and Windows Speech (open source). Voice-to-Text software systems were Dragon Naturally Speaking (commercially available) and Google Docs (open source). Average ratings for each element are listed below. Elements were scored on a 1–5 continuum where 1 = difficult, 2 = somewhat difficult, 3 = not easy but not difficult, 4 = somewhat easy, and 5 = easy.

Table 5. Evaluation components for and student rating of Text-to-Voice software systems analyzed

Text-to-Voice Software Systems	Natural Reader	Windows Speech
Learning/Understanding the System		
1. Learn system's main purpose	4.3	4.3
2. Learn system's main functions		
• Import text or document or PDF file to get speech output	4.4	3.7
• Open a webpage and get speech output	3.6	1.3
• Open an image that contains text and get speech output	3.4	1.0
• Change speaker settings	5.0	1.7
• Validate typing echo	4.1	2.0
• Use the pronunciation editor to change pronunciation of specific words	3.6	2.3
• Save speech output as an audio file	3.4	1.7
• Use basic document editing tools	4.3	4.3
3. Understand Graphic User Interface (GUI) features		
• Understand main GUI symbols	4.4	3.9
4. Understand HELP sub-system		
• Understand HELP system's features	3.5	4.2
Working with the System		
1. Main System's functions		
• Importing DOC, TXT, or PDF files into the system	4.9	3.3
• Save speech outcome as MP3 file	3.9	1.3
• Change speaker settings	4.7	2.5
• Pronunciation editor	4.1	2.3
• Spell checker functionality	4.1	4.4
• Typing echo functionality	4.1	3.0
• Inbuilt Dictionary options	3.8	4.3
• E-mail reading functionality	3.2	2.5
• Use the basic document editing tools	4.4	4.2

(*continued*)

Table 5. (*continued*)

Text-to-Voice Software Systems	Natural Reader	Windows Speech
2. Systems' Graphic User Interface GUI features		
• Navigate through system using main GUI symbols	4.0	4.2
• Change the theme of the software	2.8	3.0
• Switch to a floating bar style	3.9	2.7
• Sync option for iOS or Android devices	4.0	1.0
3. System's HELP sub-system		
• Search for a topic or question	3.3	2.7
• Navigate to web help	3.3	2.7
• Post feedback or question	3.0	2.3
Participant Opinions		
1. System useful for you?	Yes (5)	Yes (4)
	No (2)	No (3)
2. System advantageous for you?	Yes (7)	Yes (5)
		No (2)
3.Overall rating of system?	Good (7)	Good (3)
		Poor (4)
4. Preference of this system in daily life?	Often (1)	Often (0)
	Sometimes (5)	Sometimes (4)
	Never (1)	Never (3)
Participant Recommendations		
1. Any additional features recommended?	No (4)	No (1)
	Read handwriting, click in PDF, highlight as it reads, read paragraph in order	Ability to open and close webpages, ability to read images, change font when typing, make voice more natural
2. Any additional GUI features recommended?	No (5)	No (2)
	Cleaner looking icons	Easier search options, more prominent read button
3. Any additional HELP sub-system features recommended?	No (5)	No (3)
	Help function should allow questions	Tutorial on searching
4. Any recommendations to improve functionality of the system?	No (3)	No (1)
	Read text in order, click and read anywhere in document, less complicated	Make sure people know where audio button is, read emails, change reading speed, clearer access to feature.

Table 6. Evaluation Components for and Student Rating of Voice-to-Text software systems analyzed

Voice-To-Text software systems	Dragon Naturally Speaking	Google Docs
Learning/Understanding the System		
1. Learn system's main purpose	4.4	3.6
2. Learn system's main functions		
• Import an audio file or recorded speech to get output text in various formats	3.0	3.0
• Voice typing (i.e., typing text after user speaks)	3.9	4.1
• Validate spellings using voice commands	3.0	3.2
• Edit the text using voice commands	3.7	2.3
• Use basic document/text editing tools through voice input	4.0	2.4
• Use existing built-in commands and perform actions	3.7	2.8
• Train the software to understand the student accent/speech	4.3	3.2
• Create user profile	4.4	4.0
• Switch between dictation and command modes	4.6	4.0
• Customize the commands using voice	3.3	2.3
• Use the extras toolbar	3.7	4.2
3. Understand Graphic User Interface (GUI) features		
• Understand main GUI symbols	4.3	4.5
• Navigate through the system using GUI	3.7	4.0
• Change the GUI theme	3.3	2.6
4. Understand HELP sub-system		
• Understand HELP system's features	4.1	3.6
Working with the System		
1. Main System's functions		
• Use built-in commands to open any note taking application or document editor	4.1	2.8

(continued)

Table 6. (*continued*)

Voice-To-Text software systems	Dragon Naturally Speaking	Google Docs
• Use of voice typing to convert it into text	4.0	3.9
• Use basic document editing actions	3.9	4.0
• Use of spell checker using voice commands	3.4	3.8
• Change text font and size, format, and new line options through voice input	3.1	2.5
• Use built-in Dictionary options	2.3	3.2
• Save textual output in various formats (TXT, DOC, PDF) using voice commands	3.9	2.3
• Switch between dictation and command mode	4.1	3.0
• Train the software system to understand the accent, especially input nouns	4.4	2.8
• Create new or customized commands and actions associated with them	3.3	2.0
• Create User profile	4.4	3.2
• Perform system actions (i.e., open folder, play music, and close system)	3.6	3.8
• Use the extras toolbar (i.e., Transcribe recording and start playback)	3.7	3.3
2. Systems' Graphic User Interface GUI features		
• Navigate through system using main GUI symbols	4.3	4.1
• Change the GUI theme of the software	3.1	2.4
• Switch to a floating bar style/window dock style	4.3	3.6
• Look of mouse pointer	4.3	4.4
3. System's HELP sub-system		
• Search for a topic or question	4.2	3.6
• Navigate to Web help	3.8	3.8
• Post feedback or question	2.0	3.3

(*continued*)

Table 6. (*continued*)

Voice-To-Text software systems	Dragon Naturally Speaking	Google Docs
Participant Opinions		
1. System useful for you?	Yes (5)	Yes (4)
	No (2)	No (3)
2. System advantageous for you?	Yes (7)	Yes (5)
	No (0)	No (2)
3. Overall rating of system?	Poor (0)	Poor (2)
	Good (4)	Good (5)
	Excellent (3)	Excellent (0)
4. Preference of this system in daily life?	Often (1)	Often (1)
	Sometimes (4)	Sometimes (2)
	Never (2)	Never (4)
5. Rating of the quality of speech recognition	Poor (0)	Poor (1)
	Good (4)	Good (4)
	Excellent (3)	Excellent (2)
Participant Recommendations		
1. Any additional features recommended?	No (3)	No (2)
2. Any additional GUI features recommended?	No (3)	No (4)
3. Any additional HELP sub-system features recommended?	No (4)	No (3)
	Make sure microphone is close to user	
4. Any recommendations to improve functionality of the system?	Put users in a separate room so they don't look weird using software. Option to remove floating bar (3). Record teacher's lecture and read	No (2). Utilize simpler terms, more time to figure out what to say, more visible features, make available to Bradley students to put on personal computers

8 Discussion

To be successful in a university setting, students with disabilities need more supports than students without disabilities. We believe the implementation of specific software systems in Smart Classrooms within Smart Universities is a key for this to happen. College students with disabilities overall reported more positive ratings for the ability to learn and understand the main functions of the software, the ability to work with the system and the support provided them through the software for the commercially available Natural Reader text-to-voice software system as compared to the open source Windows Speech software system. In addition, participants had an overall rating of good and qualitatively wrote more positive comments about Natural Reader in that they felt the system could be useful, an advantage to use and they would use it often as compared to the Windows Speech software system. For the voice-to-text software systems, college students with disabilities also overall reported more positive ratings for the ability to learn and understand the main functions of the software, the ability to work with the system and the support provided them through the software for the commercially available Dragon Naturally Speaking as compared to the open source Google Docs software system. In addition, participants had an overall rating of good/excellent and qualitatively wrote more positive comments about Dragon Naturally Speaking in that they felt the system could be useful, an advantage to use, sometimes they would use it and the quality of speech recognition was good/excellent. Based on this data, we are suggesting that classrooms be equipped with the commercially available software Natural Reader and Dragon Naturally Speaking so that all students will have better access to the content being delivered, be able to adequately interact with the professor and classmates, and feel they are an integral part of the learning environment. In addition, these software systems also have better customer support systems for users. If this is not possible, the open source systems would be an option as all students could put the software on their own computers, but if problems arise, customer support is often lacking.

Feedback from graduate students who monitored and directed the training and testing sessions indicated that in regards to the training session, where students had a 1-h training session with presentations and demos of the software systems, the college students with disabilities did not completely know about the different types of systems, their specific functions and features and how the programs actually worked prior to the training. The students were very interested in knowing about the software because they felt it would benefit them in their ability to learn more classroom content and perform better in their classes. Most students liked the text-to-voice systems more as they felt those programs would help them in their daily classes. The main reason for this is they can obtain the voice recording of the text so that they can listen to it multiple times whenever they want. Some students mentioned that open-source tools are good as they can use them without purchasing them, but data indicated that commercially available software systems were rated higher by all students because of their features and support. Universities should investigate ways to give students with disabilities access to commercially developed systems. After working with two students in separate training and testing sessions, the format was changed for the remaining five students to one large session with training and testing combined. This change helped to ensure students attended both sessions and second, students would not forget anything from training during the delay between sessions. The students enjoyed testing rather than training as they enjoyed working directly on those systems rather than being trained on them. They also liked using their own materials to evaluate the different systems.

9   Conclusions and Future Steps

Conclusions. The performed research helped us identify new ways of thinking and our research findings enabled us to make the following conclusions:

1. Smart Universities and Smart Classrooms can significantly benefit college students with disabilities even though they are not the focus.

2. Many technologies geared towards students without disabilities will actually impact the learning of students with disabilities.
3. Some students with disabilities may need specialized technology to be successful.
4. Some technologies focusing on the success of students with disabilities may help students without disabilities to be successful.
5. For the commercially based and open-source software systems in the areas of text-to-voice and voice-to-text evaluated, the commercially based systems performed better for actual college students with disabilities.
6. More commercially based and open-source software systems should be evaluated.
7. More commercially based and open-source software systems in the areas of text-to-voice and voice-to-text should be evaluated to assess the different features and capabilities of each.
8. More research needs to be completed addressing commercially based and open-source software systems in the areas of text-to-voice and voice-to-text to decide which of them would have the most benefits for college students with disabilities.
9. More research needs to be completed with more college students with disabilities where they experience and evaluate commercially based and open-source software systems in the areas of text-to-voice and voice-to-text.
10. More research needs to be completed that directly focuses on college students with disabilities.

Next Steps. The next steps (summer 2018 – December 2019) of this multi-aspect research, design and development project deal with

1. More implementation, analysis, testing and quality assessment of text-to-voice and voice-to-text software systems with college students with disabilities.
2. Implementation, analysis, testing and quality assessment of text-to-voice and voice-to-text software systems in everyday teaching of classes in smart classrooms.
3. Implementation, analysis, testing and quality assessment of numerous components of text-to-voice and voice-to-text software systems at Bradley Hall (the home of majority of departments of the College of Liberal Arts and Sciences) and in some areas of the Bradley University campus.
4. Organization and implementation of summative and formative evaluations of local and remote college students and learners with and without disabilities, faculty and professional staff, administrators, and university visitors with a focus to collect sufficient data on quality of text-to-voice and voice-to-text software systems.
5. Creation of a clear set of recommendations (technological, structural, financial, curricula, etc.) regarding a transition of a traditional university into a smart university pertaining to software and college students with and without disabilities.

References

1. Coccoli, M., et al.: Smarter universities: a vision for the fast changing digital era. J. Vis. Lang. Comput. **25**, 103–1011 (2014)
2. Uskov, V.L, Bakken, J.P., Pandey, A., Singh, U., Yalamanchili, M., Penumatsu, A.: Smart university taxonomy: features, components, systems. In: Uskov, V.L., Howlett, R.J., Jain, L. C. (eds.) Smart Education and e-Learning 2016, pp. 3–14, June 2016. Springer (2016). 643 p., ISBN: 978-3-319-39689-7
3. Bakken, J.P., Uskov, V.L, Penumatsu, A., Doddapaneni, A.: Smart universities, smart classrooms, and students with disabilities. In: Uskov, V.L., Howlett, R.J., Jain, L.C. (eds.) Smart Education and e-Learning 2016, pp. 15–27, June 2016. Springer (2016). 643 p., ISBN: 978-3-319-39689-7
4. Bakken, J.P., Uskov, V.L, Kuppili, S.V., Uskov, A.V., Golla, N., Rayala, N.: Smart university: software systems for students with disabilities. In: Uskov, V.L., Bakken, J.P., Howlett, R.J., Jain, L.C. (eds.) Smart Universities-Concepts, Systems, and Technologies, pp. 87–128. Springer (2017). 425 p., ISBN: 978-3-319-59453-8
5. Stinson, M.S., Elliot, L.B., Kelly, R.R., Liu, Y.: Deaf and hard-of-hearing students' memory of lectures with speech-to-text and interpreting/note taking services. J. Spec. Educ. **43**(1), 52–64 (2009)
6. Ryba, K., McIvor, T., Shakir, M., Paez, D.: Liberated learning: analysis of university students' perceptions and experiences with continuous automated speech recognition. Instr. Sci. Technol. **9**(1), 1–19 (2006)
7. Nelson, L.M., Reynolds Jr., T.W.: Speech recognition, disability, and college composition. J. Postsecondary Educ. Disabil. **28**(2), 181–197 (2015)
8. Glasser, A., Kushalnagar, K., Kushalnagar, R.: Deaf, hard of hearing, and hearing perspectives on using automatic speech recognition in conversation. In: Proceedings of 19th International ACM SIGACCESS Conference on Computers and Accessibility ASSETS 2017, Baltimore, Maryland, USA, 20 October – 01 November 2017, pp. 427–432 (2017). ISBN: 978-1-4503-4926-0
9. Iglesias, A., Jiménez, J., Revuelta, P., Moreno, L.: Avoiding communication barriers in the classroom: the APEINTA project. J. Interact. Learn. Environ. **24**(4), 829–843 (2016)
10. Izzo, M.V., Yurick, A., McArrell, B.: Supported etext: Effects of text-to-speech on access and achievement for high school students with disabilities. J. Spec. Educ. Technol. **24**(3), 9–20 (2009)
11. MacArthur, C.A., Cavalier, A.R.: Dictation and speech recognition technology as test accommodations. Except. Child. **71**(1), 43–58 (2004)

Automation System of Intellectual Activity on Creating Programs in the Language of Logical Programming

Marina V. Lapenok[✉], Olga M. Patrusheva, Galina V. Pokhodzey,
Anastasiya I. Suetina, Anna M. Lozinskaya, and Irina V. Rozhina

Ural State Pedagogical University, Ekaterinburg, Russia
{lapyonok, suetina}@uspu.me, podsnejnik1993@gmail.com,
g.v.pokhodzey@mail.ru, anna-loz@yandex.ru,
irozhina@yandex.ru

Abstract. The article presents a developed software package that implements the training system that automates intellectual activity in creating programs in the language of logical programming. The choice of semantic networks as a way of the representation of knowledge in the process of modeling subject domains in the learning system during the formulation of educational logical problems is substantiated. The typology of the learning tasks used to master the logical programming language Prolog is grounded, including the following types of tasks: logical, arithmetic, value analysis tasks with subsequent selection or ordering and creation of knowledge bases on the subject domain. In this case, for each type of problem, the possibility of applying analysis of formal concepts to solve them has been investigated and a set of examples has been compiled. The technique of analysis of formal concepts and their grouping depending on the parameters of objects of the domain is developed, the essence of which is to identify an invariant set of descriptors suitable for representing different subject areas, which allows the generation of rules. The technique for generating domain representation rules based on the declaration of predicates with one or more parameters, which are correlated with the peculiarities of formulations of typical learning tasks, is developed.

Keywords: Intellectual activity · Logical programming · Prolog
Analysis of formal concepts · Rule generation · Predicate

1 Introduction

The increased demand for IT specialists in the world market is due to the fact that at present programming, including the creation of programs in the language of logical programming, is a very significant skill for the development of all spheres of the economy. In the process of training IT-specialists various teaching methods are used, including ways to increase the interest of students to learn logical programming and training tools that simulate the intellectual activity of the programmer in solving logical problems and realize the automation of creating Prolog programs.

© Springer International Publishing AG, part of Springer Nature 2019
V. L. Uskov et al. (Eds.): KES SEEL-18 2018, SIST 99, pp. 67–77, 2019.
https://doi.org/10.1007/978-3-319-92363-5_6

When studying logical programming, the trainee is offered learning tasks that describe a certain subject area in the formulation of a natural language (for example, the Russian language). The analysis of formal concepts allows you to identify the keywords in the text of the learning task and use them when generating Prolog program templates. Thus, it is possible to automate the following types of intellectual activity carried out by a person in the process of creating programs in the Prolog programming language:

- activities to analyze the text of the task with the purpose of identifying keywords;
- activities on the analysis of keywords in the formulation of the problem to determine the type of educational task and the corresponding structure of the Prolog program;
- activities on writing programs in the language of Prolog in the process of solving learning problems by generating a template Prolog program based on the identified structure.

The aim of the research was to develop a training software package that automates the process of creating Prolog programs for solving learning tasks by analyzing the text of learning tasks, identifying formal concepts in the text, determining the type of a learning task, and generating a Prolog program template for solving it.

The goal of the research was achieved through the solution of the following tasks:

- justification of the choice of the method of modeling subject domains used in the formulation of training tasks;
- development of the typification of learning tasks that can be offered to learners in the process of mastering the logical programming language Prolog;
- development of a methodology for analyzing the text of learning tasks in order to identify formal concepts (keywords) in the text and correlate them with the characteristics of typical learning tasks;
- development of a methodology for generating rules for representing a domain based on declaring predicates with one or more parameters;
- creation of a software package using the programming languages Object Pascal and Visual Prolog that automates the process of developing training programs focused on the solution of logical problems.

2   Representation of Knowledge in the Automation System of Intellectual Activity

An effective way of constructing a learning system focused on solving logical problems is to present the initial data in the knowledge bases in the form of rules for the logical derivation of relations, in the form of axioms and facts. One of the most important problems encountered in the construction of automation systems for intellectual activity is the presentation of knowledge. This is due to the fact that the chosen way of representing knowledge largely determines the characteristics of the system. The knowledge base is created for the input of knowledge, learning a system (often called acquiring knowledge) and post-processing knowledge to solve problems. The way knowledge is presented is often called the knowledge representation model [1].

When choosing the proper method of knowledge representation it is possible to avoid complicating the system and solve a lot of issues of its design and development. However, the model of knowledge representation imposes restrictions on the choice of the appropriate logical inference mechanism. That is, when designing an intelligent system, it is necessary to choose the most appropriate model of knowledge representation (and, in accordance with it, choose the logical inference mechanism used).

When analyzing possible ways of representing knowledge in learning tasks, a number of the authors justify the use of the semantic network, since the basis for this method of representing knowledge is the idea that any knowledge can be represented as a set of concepts (objects) and relations (links) [1, 2]. A semantic network is a system of knowledge that models the subject area in the form of a weighted oriented graph with a weight function and selected vertices whose nodes correspond to concepts and objects, and arcs - to relations between objects [3]. The main advantage of this model is the visibility and compliance with modern ideas about the organization of long-term human memory. The disadvantage is the difficulty of searching the output, as well as the difficulty of adjusting, that is, removing and filling the network with new knowledge.

The use of logical programming languages (e.g. Prolog), specially designed to create a mechanism of inference in intelligent systems, allows the programmer to focus on the logic of solving the problem.

3 Analysis of the Formulation of Learning Tasks to Identify Formal Concepts

In the process of developing a methodology for analyzing the text of training tasks to identify formal concepts (keywords) in the text, we used an analysis of formal concepts. The analysis of formal concepts is a method of data analysis, which involves the analysis of the following sets: objects, descriptors (characteristics of objects) and relations (attributes of objects due to the inherent object characteristics) [4]. To increase the effectiveness of creating training software packages oriented to solving logical problems, it is advisable to develop a methodology for analyzing formal concepts and grouping them depending on the parameters of domain objects, the application of which will make it possible to justify algorithms for generating domain representation rules in software complexes.

The method of the analysis of formal concepts and their grouping depending on the parameters of objects in the domain is based on the identification and justification of an invariant set of descriptors (characteristics) suitable for representing different subject areas [5]. For example, such characteristics as 'condition', 'restriction', 'conformity', 'expression', etc. Based on the analysis of the characteristics of typical learning tasks and subject areas, four sets of descriptors corresponding to the main types of learning tasks were identified: tasks for performing arithmetic calculations, logical tasks, tasks for analyzing values with subsequent selection or ordering, tasks for creating knowledge bases on the subject domain. The following agreements were adopted in the study. The educational task was assigned to the first type of tasks – the type 'for performing arithmetic calculations', if in the process of analyzing the text of this learning task the following descriptors were identified in the formulation: 'calculate' (or synonyms 'calculate', 'determine', 'find the value', 'expression value', 'whether ...') and 'formula' (or synonyms 'analytic expression', 'variable name', 'function name (parameters) ='). If in the formulation of the problem there were identified descriptors 'set compliance' (or synonyms 'distribute by places', 'find the corresponding pairs of elements') and 'constraints' (or synonyms 'if <condition>', 'facts', 'contra-indications', 'self-restraint'), then such a task was assigned to the second type - the type of 'logical tasks'. If the goal of the problem is to find a local extremum (extreme elements of the series), which is expressed in descriptors 'most ...', 'find the smallest element', 'set order' and 'comparison' (presented in the form of alternatives 'maximum/minimum', 'above/below', 'more/less', 'older/younger', 'left/right'), then this task was assigned to the third type - the type of 'task for the analysis of values with subsequent selection or ordering'. The educational task was related to the fourth type - the type of 'task for creating knowledge bases on the subject area', if in the process of analyzing the text in the formulation of the task such descriptors as 'create' (or synonyms 'realize', 'form', 'develop') and 'knowledge base' (or synonyms 'directory', 'project', 'database', 'information') were identified.

4 Analysis of Formal Concepts and Generation of Rules of Presentating of the Subject Domain in Software Complexes

The methodology for generating rules for representing a domain is based on declaring predicates with one or more parameters that are correlated with the characteristics of the formulations of typical learning tasks.

In problems of the first type 'for performing arithmetic calculations', the target characteristic (that is, a numeric variable whose value determines the truth conditions of the target predicate) is a set of predicates {<write ()> , <read ()>} that reach the value 'truth' in the calculation formula and output the solution to the screen. To solve 'logical problems' it is necessary to set a correspondence between objects, i.e. define in the program several groups of single variables (or several arrays of variables), considering the above variables as elements of finite sets, between which one-to-one correspondences are established. The number of elements in the corresponding sets should be the same [6]. In 'problems for the analysis of values with subsequent selection or ordering' the target characteristic (i.e. elements whose sequence determines the truth conditions of the target predicate) is a predicate <series ()> that reaches the value of 'truth' when finding the desired element in the resulting sequence and output the result to the screen. In 'problems of creating knowledge bases on the subject domain', the target characteristic (i.e. the fact base including composite objects) is a set of statements of the target predicate that reaches the value of 'true' when sampling, searching by criteria or outputting the entire database to the screen.

In the process of the research, a program complex was developed using the programming languages Object Pascal and Visual Prolog, which implements the graphical interface [2] and the generation function of the standard Prolog program template, designed for mastering logical programming, which in turn is necessary for solving logical problems. Activating the objects (buttons) placed on the main page of the developed application (Fig. 1) initiates the execution of the examples of the Prolog programs corresponding to the typical tasks.

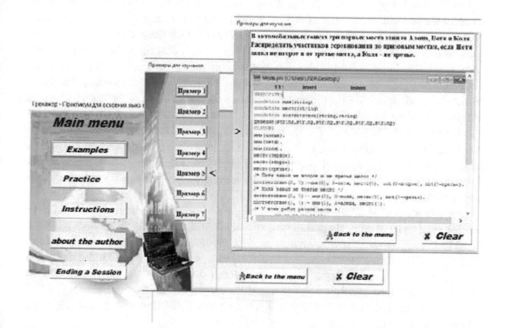

Fig. 1. Interface of the simulator-practice

After studying the standard examples, the services are available that allow you to enter the wording of the problem in text format, identify the type of the problem by analyzing formal concepts (i.e., by identifying descriptors in the text of the task) and generating the corresponding Prolog program template.

In the Prolog program template, the description of the subject area is presented in the form of a set of facts (based on predicates of various types), as well as a set of rules constructed in accordance with the developed methodology.

For tasks of the first type 'to perform arithmetic calculations' in the template, the target characteristic is described by a predicate with parameters. The essence of the function, as well as the constraints imposed on the domain of the definition, is described in the rule of the function. For example, if there are descriptors to 'calculate' and 'value of an expression' in the formulation of the task, then in the **Predicates** section you should specify the type of predicate objects in the form: <*value (real, real*)>; in the **Clauses** section, you should describe a rule that includes the essence of the function, as well as restrictions imposed on the scope of the definition. In the **Goal** section, you should describe the target predicate, which includes the subgoal of typing numerical values using <*write ()/read ()*> predicates, as well as the subgoal that implements the predicate of calculating a given function and outputting the solution to the screen.

When solving 'logical problems' in the Prolog program, finite sets should be described as a set of facts, and the dependencies between objects are set using rules between the values of which one-to-one correspondences are established. For example, if there is a descriptor 'distribute by places' in the formulation of the task, then in the **Predicates** section, you must specify the predicate name and type of the predicate objects in the form: <*match (string, string)*>. This description means that both predicate objects <*match ()*> are strings. The **Clauses** section describes the facts (each fact ends with a 'dot' symbol) and rules that include constraints imposed on the scope of the definition, as well as the target predicate <*solution ()*>, which determines the correspondence task specified by the formulation. In the **Goal** section the target predicate and the predicate <*write ()*> are accessed to display the solution on the screen.

When solving 'problems for the analysis of values with subsequent selection or ordering' to find combinations and/or permutations caused by the content of the learning task and the rules of combinatorial analysis, the program lists the facts and describes the rules, and then declares the predicate, representing a sequence of matches and expressing the semantic solution of the problem. For example, if there are descriptors 'define' and 'high/low' in the formulation of the task, then in the **Predicates** section, you must specify the string type (domain) of the predicate objects in the form: <*string (string, string, string, string, string, string)*>. The **Clauses** section should describe facts and rules that include constraints imposed on the scope of the definition, as well as the target predicate <*series ()*>, which defines the 'highest' and 'lowest' object. In the **Goal** section, using the predicates of <*write ()*> and <*read ()*> it is necessary to provide keyboard input of numeric variables that will access the target predicate and obtain the result on the screen.

In tasks of the fourth type 'Creating knowledge bases on the subject domain' in the Prolog program, the target characteristic is described by a predicate with parameters representing composite objects. Knowledge (i.e., true statements about the subject domain) is described as a set of facts, and the goal is the predicate <*write ()*>, which implements the output of the database to the screen and other queries (search by criteria). For example, if there are descriptors of the form 'create', 'project', 'implement', 'directory', 'contain' and 'information' in the formulation of the task, then in the program **Domains** section you should create user domains of objects from the basic domain types in the form: <*domain name = name of the base domain*>. In the **Predicates** section, you specify the type (domain) of the predicate objects in the form: <*name_of_the_predicate (domain_name1, domain_name2)*>, where *domain_name1* and *domain_name2* will mean certain sets of values. In the **Clauses** section it is necessary to describe the facts (logical assertions) that are derived from the same target predicate name, and in the **Goal** section it is necessary to access the target predicate and the <*write () and read ()*> predicates to output the knowledge base to the screen and organize the search by the value you type.

Below there are the examples of the application of the developed method for solving typical educational tasks.

Example 1. The formulation of the task. Create a project that implements the railway directory. The directory contains the following information about each train: train number, destination and departure time. Output all the information from the directory and organize a search for the train by destination.

Solution. Since there are descriptors 'create', 'project', 'implement', 'reference', 'is contained' and 'information' in the task definition, it is possible to define the task type as 'tasks for creating knowledge bases by the subject domain'.

In the **Domains** section, we create our own types of objects from the basic domain types, as follows: number = integer and destination, time = string (which refer to the integer and string type, respectively). The **Predicates** section specifies the type of the predicate objects in the form: *<train (number, destination, time)>*. The **Clauses** section describes the facts (statements). In other words, all the statements are variations of the same target predicate *<train ()>*. Finally, in the **Goal** section, you access the target predicate and the predicate *<write ()>* to display the knowledge base on the screen and the predicate *<read ()>* to organize the search for the train on the destination entered from the keyboard. Thus, the Prolog program template will look like this:

```
DOMAINS
name1= domain_type1
name2= domain_type2
name3 = domain_type3
...
PREDICATES
nondeterm knowledge_base (name1, name2, name3,...)
CLAUSES
knowledge_base (value11, value12, value13,...).
knowledge_base (value21, value22, value23,...).
knowledge_base (value31, value32, value33,...).
...
GOAL // output the entire database
write("    <Title of the database>    "), nl,
write("name1    name2    name3    ... "),
nl, knowledge_base (A,B,C,...), write(A," ",B,"   ",C," ",...), nl, fail.
```

Next, the user (a student studying logical programming) needs to edit the template in accordance with the formulation and condition of the task.

Example 2. The formulation of the task. In the car race, the first three places were taken by Alex, Pete and Nick. Distribute the participants of the competition by prize-winning places, if Pete took neither the second nor third place, and Nick didn't take the third one.

Solution. Since there is a descriptor 'distribute by places' in the formulation of the task, the task is logical. In accordance with the methodology for generating domain presentation rules, the Prolog program template must include a description of the predicates in the **Predicates** section in the form: *<match (string, string)>*. This description means that both predicate objects *<match ()>* are strings. In the **Clauses** section, the template must contain a description of the facts and rules that include the constraints imposed on the scope of the definition, and the target predicate *<solution()>* that determines the correct matches. In the **Goal** section of the Prolog program template, you must access the target predicate and the *write()* predicate to display the solution on the screen. From this it follows that the Prolog program template will have a predefined structure:

PREDICATES
nondeterm predicate_name1 (predicate_object_type1)
nondeterm predicate_name2 (predicate_object_type2)
...
nondeterm match (predicate_object_type1, predicate_object_type2,...)
solution (predicate_object_type1, predicate_object_type2,
predicate_object_type1, predicate_object_type2, predicate_object_type1,
predicate_object_type2,...)
CLAUSES
predicate_name1 (predicate_object_name1).
predicate_name1 (predicate_object_name2).
...
predicate_name2 (predicate_object_name1).
predicate_name2 (predicate_object_name2).
...
match (X, Y):- < Condition 1>.
match (X, Y):- < Condition 2>.
...
match (X, Y):- < Condition n>.
solution (X1,Y1,X2,Y2,...,Xn,Yn):-
X1= predicate_object_name1, match (X1,Y1),
X2= predicate_object_name2, match (X2,Y2),
...
Xn= predicate_object_name n, match (Xn,Yn),
Y1<>Y2, Y2<>Yn, Y1<>Yn, ...
GOAL
solution (X1,Y1,X2,Y2,...,Xn,Yn),
write(X1," - ",Y1,...," - ",Z1),nl,
write(X2," - ",Y2,...," - ",Z2),nl,
...
write(Xn," - ",Yn,...," - ",Zn),nl.

Next, the student needs to edit the template in accordance with the formulation and condition of the task.

Example 3. The formulation of the task. Determine the highest and lowest tree, if it is known that the poplar is higher than the birch, which is higher than the linden. The maple is lower than the linden, and the pine is higher than the poplar and lower than the spruce.

Solution. Since there is a descriptor 'define' and 'high/low' in the formulation of the task, it is necessary to define the type of the problem as 'tasks for the analysis of values with subsequent selection or ordering'. The **Predicates** section specifies the type of the predicate objects in the form: <*series (string, string, string, string, string,*

string)>. The **Clauses** section describes the facts and rules that include constraints imposed on the scope of the definition, as well as the target predicate *<series ()>*, which allows you to determine the highest and lowest tree. In the **Goal** section, using the *<write ()>* and *<read ()>* predicates, the keyboard input of numeric variables is performed, after which the target predicate is accessed and the result is displayed on the screen. This implies that the Prolog program template will have a predefined structure:

```
PREDICATES
nondeterm predicate_name (predicate_object_type1, predicate_object_type2)
series (predicate_object_type1, predicate_object_type2,
predicate_object_type1, predicate_object_type2,
predicate_object_type1, predicate_object_type2)
CLAUSES


predicate_name (value11, value12).
predicate_name (value21, value22).
…
predicate_name (valueN1, valueN2).
series(X1,X2,…,Xn):-
predicate_name (X1,X2), predicate_name (X2,X3),…, predicate_name (Xn-1,
Xn).
GOAL
series(X,Y ,…,Z),write(X,"-",Y,"-",…, "-",Z), nl.
```

5 Software Complex 'Simulator-Practice for Mastering the Programming Language Prolog'

The software complex 'Simulator-practice for mastering the programming language Prolog' provides conditions (by providing program templates) for program coding in the logical language for implementing learning tasks of various types (including types: logical tasks, arithmetic calculations, database knowledge, etc.). In this case, an automated analysis of the text formulation of the task is performed, keywords are highlighted, on the basis of which a template of a certain type is formed. In the template, the description of the domain is represented in the form of a set of facts (based on predicates of a certain arity), and also in the form of a set of rules constructed in accordance with the developed methodology.

The program for automating the analysis of formal concepts with the selection of keywords and generation of standard Prolog templates was coded in the programming language Object Pascal in the Delphi programming environment. The development of the program template was performed in the Visual Prolog programming environment with the ability to create a visual part of the program (user interface). The following data formats are installed: the input data are the text of the task to be implemented, the output data are the generated program template, suitable for solving a specific task. After launching the application, a drop-down menu appears on the welcome form that allows you to start working with the program (Fig. 1).

The first block provides examples of tasks that demonstrate the structure of the complete program, which allow studying typical educational tasks for further training in logical programming. In the second block, a set of tasks for independent work is compiled, which are analyzed by the program for generating a template. The task text can be entered into a special window on the form from the keyboard, after analyzing the text formulation of the task, the type of the task is displayed on the screen and an appropriate template is generated. When you click the Start Prolog button, the Visual Prolog development tool, which opens the generated template, starts and the task is directly resolved.

The testing of the programme complex was attended by the students of the Ural State Pedagogical University, who evaluated the questionnaire process:

- structure of the training seminar;
- visual design of the application;
- usability of the application;
- availability of the presentation of the training material in the examples proposed in the simulator;
- possibility of self-mastering the course of logical programming with the help of the developed training simulator.

In the viewpoint of the students, the 'Simulator-Practicum' application facilitates to master the logical programming of the Prolog.

6   Conclusion

As the result of the research, the following tasks were solved.

The choice of semantic networks as a way of representing knowledge in the modeling of subject areas in the training system that automates intellectual activity in the development of programs in the language of logic programming Prolog for solving logical problems is substantiated.

The typification of the learning tasks used to master the logical programming language Prolog is developed, including the following types of tasks: logical, arithmetic, value analysis tasks with subsequent selection or ordering, and the creation of knowledge bases on the subject domain. In this case, for each type of problem, the possibility of applying analysis of formal concepts to solve them is investigated and a set of examples is compiled.

The methodology for analyzing formal concepts and their grouping depending on the parameters of objects in the domain is developed, the essence of which is to identify an invariant set of descriptors (characteristics) suitable for representing different subject areas, which allows the generation of rules.

The methodology for generating rules for representing a domain is developed, based on the declaration of predicates with one or more parameters, which are correlated with the peculiarities of formulations of typical teaching tasks.

The implemented training system automates intellectual activity in the development of programs in the language of logical programming. Prolog is a program complex in the programming languages Object Pascal and Visual Prolog). The approbation confirmed the feasibility of the applications developed for teaching students to logical programming, in particular, the compilation of Prolog programs with the structure most suitable for solving typical tasks.

References

1. Cussens, J.: Issues in learning language in logic. In: Proceeding Computational Logic: Logic Programming and Beyond, Essays in Honour of Robert A. Kowalski, Part II, pp. 491–505. Springer-Verlag, London, UK (2002)
2. Borovskaya, E.V., Davydova, N.A.: Fundamentals of artificial intelligence/Training manual, BINOM. Laboratory of Knowledge, p. 127 (2010)
3. Yasnitsky, L.N.: Intellectual systems/tutorial. Laboratory of Knowledge, 221 p. (2016)
4. Davydova, E.A., Lapenok, M.V.: The Application of the analysis of formal concepts for the generation of the rules of the representation of a subject domain when creating educational software/Actual problems of teaching mathematics, informatics and information technologies, no. 2, 170–181 pp. (2017)
5. Kuznetsov, S.O.: Lattices of formal concepts in modern methods of analysis and data development, 86 p. Pospelov Readings (2011)
6. Bratko, I.: Programming in the language PROLOG for artificial intelligence, 243 p. Mir, Moscow (1990)

Historical, Cultural, Didactic and Technical Challenges in Designing a Learning Environment for the Protestant Reformation on a Large Touch-Display

Gudrun Görlitz[✉]

Department of Computer Science and Media,
Beuth University of Applied Sciences, Berlin, Germany
goerlitz@beuth-hochschule.de

Abstract. The paper introduces a gesture-driven, complex web application that conveys historical content for students and tourists on the effects of the Protestant Reformation on a 100 in. touch-display inside the Historical Superintendency in Torgau. Seven restored historic buildings have been modularly implemented as interactive 360° panoramas that are "walkable" within the application and serve as an anchor point for interactions. Each module comprises a set of mini-games and interactions that allow students and tourists to explore and interact with historical-theological content. The article presents historical, didactic and technical challenges to design such a cross-cultural learning environment.

Keywords: Informal learning · Historical content · Hardware
Software · 3D panorama · Learning and teaching activities
Gesture-driven web application

1 Introduction

The German town Torgau played an important role in the German reformation between 1517 and 1648. Many tourists travel to the town to gain knowledge about the historical buildings and the personalities of the reformation, such as Luther [1].

Funded by the European Fund for Regional Development (ERDF) the Beuth University of Applied Science in Berlin develops interactive applications for touch modules. Within this project the University, in cooperation with the protestant youth education project Wintergrüne in Torgau, has developed a complex game based web-application to convey historical content.

2 Research Project Objectives

The aim of the project was to develop an application for a 100 in. touch-display with which students and visitors could learn about the history of the Protestant Reformation. The application is displayed in the Historical Superintendency in Torgau, where Martin

© Springer International Publishing AG, part of Springer Nature 2019
V. L. Uskov et al. (Eds.): KES SEEL-18 2018, SIST 99, pp. 78–83, 2019.
https://doi.org/10.1007/978-3-319-92363-5_7

Luther, Philipp Melanchthon, Justus Jonas and Johannes Bugenhagen created theological reports in 1530, which are called the "Torgauer Artikel" [2]. Because many of the restored old buildings in Torgau played an important role in the Reformation, it was decided to connect the content to be conveyed with the buildings. The following overview shows the historical content and the building, where it is located (Table 1):

Table 1. Overview of the modular structure of the application

Building	Historical Content
Johann-Walter-School with church	The Protestant Reformation and its impact on the educational system
Spalatin house	World religions Important representatives of the Protestant Reformation
Castle Hartenfels	Relations between church and state Residence of the Prince-Electors
Chapel of the castle	Differentiation of Protestantism and Catholicism
Marien-church	"Fourteen Holy Helper" Cranach-painting, Music in church
Katharina von Bora's Death House	Family life in the late medieval/in the early Modern times The role of a woman
Historical Superintendency	New administrative structures of the Protestant church
Town hall	Tasks of the city council

3 Background and Related Work

Museums around the world have expanded their exhibitions to include sophisticated, cost-intensive multimedia installations that convey historical content to visitors. (see for example [3–5]) So "the museum provides opportunities for visitors to personalise their use of and engagement with collections. This user-focused approach allows visitors to construct their own meanings, using collections in creative and non-traditional ways" ([5], p. 130).

The challenge is that historians, museum educators and IT specialists work very closely together to ensure that the multimedia exhibits achieve a good quality.

4 Media-Didactical Concept

The basis of the didactical concept is a user-centered design. The application should have a modular structure. Each building with its explorative embedded interactions and mini-games is a module. The application can be extended by further modules. There should be no dependency between the mini-games and interactions. The user can activate the mini-games in any order. The interactions and mini-games in a module do not have to be completely processed. It should be easy to switch between the modules.

The modules are accessible through the main interface, that shows seven buttons in the form of historical buildings to start the application (see Fig. 1).

Fig. 1. Start-screen with the historical buildings in Torgau

19 different 3D-panoramas of the most important historical buildings of Torgau have been created to realize an immersive interface. The idea of users looking into the past and dealing with the impact of the Protestant Reformation on today's society comes to life. Interactive 2D-elements are integrated into these panoramas and allow for explorative learning. A modern screen-design targets the predominantly young audience. Gamification-elements [6] serve as a template for the engagement of young adults.

5   Example: Johann-Walter-School with Church

The students of the Johann-Walter-High-School in Torgau have their auditorium and the music cabinet in an old church, in the past named "Alltagskirche (everyday church)". Figures 2 and 3 visualize two 3D inside panoramas of the Alltagskirche with the integrated interaction buttons for the mini-games.

The module comprises a set of mini-games that allow students and tourists to explore and interact with historical-theological content. In the "Alltagskirche" were implemented interaction around the school to learn about the Protestant Reformation and its impact on the educational system. Four of the mini-games are shown as examples in the following Table 2:

An animated encyclopedia (short texts, audio, images and videos) of 76 pages summarizes the information contained within each mini-game. The application for the multimedia touch-display is intended for guided workshops with students and tourists.

Fig. 2. 3D inside panorama of the Alltagskirche

Fig. 3. 3D inside panorama of the music cabinet in the Alltagskirche

6   Implementation

The research team decided to use 3D panoramas as the main user interface, since many historical buildings were restored. The JavaScript library Krpano [7] offers a wide range of tools that turn panoramas into virtual tours. Custom actions like hotspots can be added to the panorama to alter the view or add areas to interact with.

Table 2. Examples of mini-games oft he modul Alltagskirche

Name of the mini-game	Example screenshot
School life in woodblock printing	
School rules in the past and today	
Monk memory	
Latin exercise	

The application was developed with HTML5, CSS3 and JavaScript and optimized to run in Google Chrome browser. The event handling was implemented with jQuery, pictures and interface elements were animated with pep.js (drag gestures) and Hammer. js (touch gestures), both JavaScript libraries.

7 Summary and Conclusions

In total 40 games and interactions controlled by touch gestures were developed. The application has already been installed in the Historical Superintendency in the old Torgau.

This project shows the large potential of E-learning technologies working collaboratively with church historians and technology researchers. The modularity of the application as well as the variety of different interactions and mini-games enable the project partner to plan and coordinate workshops according to the needs and expectations of the visitors.

Direct interaction with the touch-display goes hand in hand with group discussions, since visitors often arrive as groups and have a lively exchange of preexisting knowledge or recently received information. Recent feedback from teachers and students shows that the application performs very well for learning purposes. Teachers are able to use some of the applications content for their teaching content as students received a new method to internalize new information.

References

1. Schilling, H.: Martin Luther. Verlag C. H. Beck (2014)
2. Decot, R.: Torgauer Artikel. In: Kasper, W. (Hrsg.) Lexikon für Theologie und Kirche. 3. Auflage. Band10. Herder, Freiburg im Breisgau (2001)
3. Trendswatch 2012: Museum and the Pulse of the Future. American Association of Museums (2012)
4. Eisele, B.: Révolution Francaise: der Louvre definiert den Multimediaguide neu. In: Sieck, J., Franken-Wedelstorf, R. (Hrsg.) Kultur und Informatik: Aus der Vergangenheit in die Zukunft, pp. 49–57. VWH-Verlag (2012)
5. Clifford, P.: Blended Learning in the cultural sector. Where we can go from here? In: Sieck, J. (Hrsg.) Kultur und Informatik: Multimediale Systeme, pp. 129–131. VWH-Verlag (2012)
6. Zichermann, G., Cunningham, C.: Gamification by Design. O'Reilly, Sebastopol (2011)
7. krpano Panorama Viewer. http://www.krpano.com. Accessed 21 Jan 2018

Intellectualization of Educational Information Systems Based on Adaptive Semantic Models

Tamara Shikhnabieva[1(✉)], Alexey Brezhnev[2],
Marida Saidakhmedova[2], Aleksandra Brezhneva[2],
and Seda Khachaturova[2]

[1] Plekhanov Russian of University Economics, The Federal State Budget
Scientific Institution "Institute of Education Management of the Russian
Academy of Education", Moscow, Russian Federation
shetoma@mail.ru
[2] Plekhanov Russian of University Economics, Moscow, Russian Federation
brezhnev.AV@rea.ru, msaidaxmedova@mail.ru,
seda_@mail.ru, anbrl@rambler.ru

Abstract. For effective functioning of information systems and technological processes in the field of education, as well as strategic management of educational institutions, it is necessary to solve a number of problems. The paper presents a brief analysis of the results of the author's research team on the problems of intellectualization of information systems and technological processes in the field of education, as well as the proposed approaches to solve them. In the existing intellectual information systems there is no their purposeful use for management of educational process according to the required principles of didactic systems. In addition, the problem of realization of interactivity and creativity in teaching with the use of intellectual information systems for educational purposes (IISEP) and determining its quality is insufficiently investigated. To eliminate the existing shortcomings of IISEP, we propose to create intelligent modules that will be integrated directly into IISEP.

Keywords: Intellectual information systems · Education
Knowledge representation · e-Learning · Adaptive semantic models

1 Introduction

To implement the tasks facing modern education, we need an efficient, flexible, modular system based on t he most advanced technologies and training facilities.

A distinctive feature of the modern stage of education is the search of teachers-researchers ways to use formal methods to describe the learning process using the apparatus of system analysis, Cybernetics, synergetics, taking into account the development and expansion of concepts, principles and achievements of didactics.

At the present stage of the evolution of the educational system closer to the threshold beyond which we should expect massive use of semantic technologies and intelligent information systems for education (further – IISE).

V. L. Uskov et al. (Eds.): KES SEEL-18 2018, SIST 99, pp. 84–93, 2019.
https://doi.org/10.1007/978-3-319-92363-5_8

Intellectualization of information systems for educational purposes is also due to the need to introduce high-tech automated information systems not only in the learning process, but also directly into the management of educational institutions in order to improve its quality.

Our article is devoted to generalization and systematization of experience of use of IISE, and also to identification of problems and perspective directions in their creation and development in relation to conditions of domestic education system.

In addition, the paper presents a detailed description of our developed intelligent systems learning and knowledge control based on adaptive semantic models, which is used in a number of universities in distance learning.

2 The Intellectualization of Educational Information Systems: A Literature Review

Analysis of the use of information systems of automation of educational process in educational institutions revealed a number of inconsistencies that violate the two main tendencies of modern education – the differentiation and integration.

As the review and research of works on intellectualization of educational information systems has shown, over the last decade a number of directions have been formed [1–5, 7].

These areas are associated with the development of various types of intelligent information systems for educational purposes. However, the problem of implementation of interactivity and creativity in learning using intelligent systems and determining its quality is not sufficiently investigated [2, 4, 8–10].

Also in the existing intellectual information systems there is no their purposeful use for management of educational process according to the required principles of didactic systems [6]. For the effective functioning of intellectual information systems for educational purposes, it is necessary to solve a number of problems related to the presentation, processing, storage and use of knowledge [3, 4, 9–12].

These include:

- development of methods, models of knowledge representation and structuring in educational systems;
- improvement of management of educational process of educational institutions on the basis of model representations of knowledge base of educational appointment;
- improving the efficiency of the educational process, optimizing and improving the quality of the management process in educational institutions;
- development of new approaches to the formalization of knowledge using different methods for the intellectualization of the information environment of the educational institution [13];
- development of methods of information educational environment formation based on the use of modern information systems, artificial intelligence technologies.

Some problems are related to the representation, processing, storage and use of knowledge in intelligent systems of educational purpose (Fig. 1) [7]. Representation of

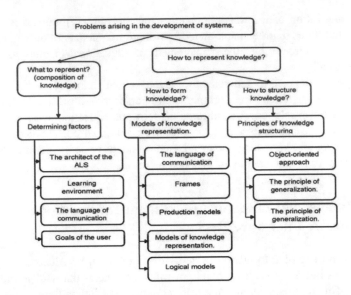

Fig. 1. The problems that arise when developing systems based on knowledge.

knowledge in information systems for the purpose of their intellectualization and abstraction occurs in almost all spheres of activity, not only in education.

Consider the intellectualization of information systems in education in terms of the use of certain technologies that allow them to create. These are agent-oriented technologies, expert systems technology, artificial neural networks, fuzzy logic, genetic algorithms and a number of others. In intellectual information systems of educational appointment these technologies are guided by the modern organization of education at all its levels.

One of the types of intelligent systems is integrated expert systems [15]. As an instrumental means of building such intelligent systems, including network adaptive intelligent learning systems, the complex at-TECHNOLOGY developed on the basis of the laboratory "Intelligent systems and technologies" of the Department of "Cybernetics" of the national research nuclear University MEPhI is of interest. As the author [6] points out, the effective functioning of these network adaptive intelligent learning systems is supported by the system dynamic modification of software "with the help of a set of unified procedures" on the basis of the current working version of the complex. Modification of software includes modification of implemented algorithms, models and methods, as well as source code and script, including the relationship of the modified component with other software. As part of the training courses in this area of study, the issues of heuristic models of knowledge representation are studied. Here, special attention is paid to network models of knowledge representation, including modeling of the simplest situations of the problem area with the help of frame models of knowledge description and semantic networks. Currently, as part of the tools to support the construction of the learning model in the network version of the complex IT-TECHNOLOGY, there are a number of components to identify the abilities of the student to simulate the simplest situations of the subject area with the help of frames and semantic networks. Depending on the type of the model of the student and his/her individual approaches to learning (in General, approaches can be inductive, deductive and hybrid), it is proposed to use three learning vectors (fast, normal and slow) [13].

On this basis, the process of real learning is imitated taking into account its characteristic features, such as the mutual integration of the processes of verification of the models of the student, the teacher and the training course, the ability of the student, the optimality of the strategy of dosing knowledge and exercises by the teacher, the speed of memorizing and forgetting knowledge by the student, the duration and stability of its active state, etc. Recently, adaptive hypermedia systems and multi-agent intelligent systems are also used, in the design of which an agent-oriented approach based on the use of intelligent agents is implemented [16, 17].

Agent-oriented approach to building intelligent information systems for educational purposes involves the creation of an information agent-oriented training complexes using virtual worlds subject areas is a complex dynamic models of the subject areas that come closest to the reality of simulated environments and situations [17].

The architecture of such complexes is a complex distributed multi-user information system, which can be adequately interpreted using multi-level architectural models.

Each level is relatively independent, can be described separately and developed autonomously, and the ways of interaction between the levels are unified.

Thus, the analysis of existing intellectual information systems of educational purpose showed that not enough attention is paid to the development of systems integrated into the network information and educational environment [6, 7, 14]. Also, our research has shown that it is necessary to improve the methods and models of knowledge representation in IISEP, in order to their effective functioning.

3 Our Approaches to Improvement of Information Systems of Educational Appointment

In the course of researches on the basis of monitoring of the modern state of use of intellectual information systems of educational appointment, new approaches to elimination of shortcomings of the existing intellectual information systems are proved and offered.

One of the ways to solve these problems is the creation of intelligent modules that will be integrated directly into the IISEP, which will provide: building a sequence of individual training courses; intellectual analysis of the answers of students; interactive support in solving problems [18].

In our opinion, the perspective direction of improving the University management system is to build a model of school management using intellectual methods and models.

Studies have shown that this is a kind of metamodel, which forms the basis of the educational knowledge base and has a three-level hierarchical structure [18].

At the first level of the hierarchical model there are generalized knowledge about the processes of learning a particular area of training or specialty. Initial material for model representations are the state educational standards; standard (approximate) curricula.

On the basis of a comparative analysis of intelligent methods and models, as foundations of the model of representation of knowledge of school, the selected multi-level adaptive hierarchical semantic network.

However, it should be noted that frames, production models and others can be used as the basis of the model, as well as their combination.

The second level of metamodel is designed to represent the knowledge of specific disciplines studied in the process of training, and details the knowledge of the first level. The main sources of informative information of this level are: data of educational programs of the disciplines reflecting structure of educational material; the recommended educational and methodical literature; materials of scientific and methodical conferences and exhibitions providing the advancing training of students.

The basis of the third level of metamodel is also an adaptive semantic network that reflects the logical structure of the educational material and shows the causal relationship between its concepts. The source data for the model presentation at this level are thematic plans of academic disciplines and professional knowledge of teachers.

On the basis of the analysis of the existing intellectual information systems of educational appointment and methodical approaches to their construction, we offer theoretical and methodical approaches to structuring and logical representation of knowledge in IISEP based on adaptive semantic models [7].

In justifying the choice of the model of knowledge representation in IISEP, we assumed that learning is a kind of cognitive process that takes place in specific conditions, and involves the interaction of the teacher, the student, the object of knowledge and the phenomena of In addition, the problems we consider are related to the processing of semantic information expressed by signs and the presence of subjective factors.

In addition, the problems we consider are related to the processing of semantic information expressed by signs and the presence of subjective factors [19]. There are many ways of presenting semantic information, one of which is the semantic network.

When solving creative problems, which include the learning process, requires a system of knowledge representation based on the logical-semantic approach. Such a system allows to display the conditions of the problem in the form of a structured model, which takes into account all the necessary connections between the elements.

It is known that a structurally semantic network is an oriented graph whose vertices denote concepts and arcs represent relations between them.

As is known, "the graph is a very characteristic mathematical object of adaptation" [20]. The standard objective of learning is usually to learning the best way to remember certain pieces of information (lexical units). The effectiveness of Q such training can be assessed by the results of periodic control of the student. The efficiency depends on the learning algorithm of U and the disciple [20]:

$$Q = Q(U, \omega) \tag{1}$$

where ω is the individual characteristics of the pupil as an object of study. It is obvious that these properties are a priori unknown explicitly and can be obtained only as a result of a rather cumbersome identification process.

The learning process is natural to make adaptive, i.e. adapting to the individual characteristics of the student, which, generally speaking, can change in the learning process, i.e.

$$\omega = \omega(t) \tag{2}$$

can be done by appropriate selection of the pieces of information U for learning, i.e., solving the problem of adaptation.

In addition, the construction of the theory of learning, reflecting the main aspects of reality and providing opportunities for improving the learning process, based on the appropriate model of learning activities. Well-known models in the form of graphs, matrices, logical equations, predicates, probabilistic and deterministic automata are not suitable for describing the learning process, as well as for solving problems of education management since they are more focused on the analysis and generalization of quantitative information. In this case, it is necessary to take into account subjective factors, specifics of semantic information when choosing a method of its presentation, processing, analysis and dynamics of intensive development of a number of subject areas.

Great potential to solve these problems, we have adaptive semantic model [7, 19].

Adaptive semantic model of educational material is a multilevel hierarchical structure in the form of semantic network represented by oriented graph, in the vertices of which are the concepts of the studied subject area, and the edges indicate the connection between of them.

Figure 2 shows an example of a single-level adaptive semantic model that explains the purpose and properties of the model.

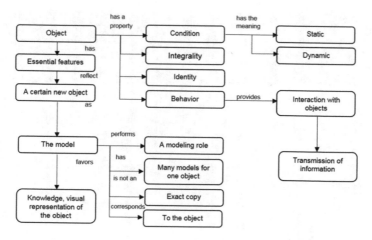

Fig. 2. The semantic model on the topic "model Properties".

The advantage of semantic networks as a model of knowledge representation and the learning process itself is the visibility of the description of the subject area, flexibility, adaptability to the purpose of the student. However, the property of visibility with the increase in size and complexity of the links of the knowledge base of the subject area is lost. In addition, there are significant difficulties in handling various types of exceptions. To overcome these problems, the proposed method is a hierarchical description of networks - allocation in local subnets located at different levels.

It should be noted that the use of semantic networks to represent the knowledge of the subject area also allows to organize information retrieval mode, in which there is a request for information depending on the educational situation.

4   Result of Investigation

On the basis of the offered theoretical and methodical approaches we have developed the intellectual information system of training and control of knowledge based on adaptive semantic models [7, 19] which is used in training in a number of educational institutions in the remote form. This section presents the principles of construction and experience of using the developed intellectual system in training.

On the basis of the offered theoretical and methodical approaches we have developed the intellectual information system of training and control of knowledge based on adaptive semantic models which is used in training in a number of educational institutions in the remote form. This section presents the principles of construction and experience of using the developed intellectual system in training.

Working with the automated training system "CASPIAN" begins with the launch of the executable Teach file.exe. The main window of the program has the view shown in Fig. 3.

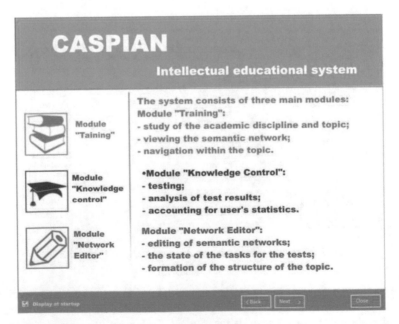

Fig. 3. The main interface of the intellectual educational system "CASPIAN".

The system has a modular structure, implemented in an object-oriented program-ming environment. Press the "Next" control button in the Fig. 3, opens the main form of the program, consisting of two pages: Content and User. On the Contents page is displayed, the navigation history, user settings, a list of items of the knowledge base, the control panel are displayed in a list and push button start system. The main component of the page is a list of knowledge base elements. As an element of the base can be the direction of training (specialty), subject, theme and educational network. As an element of the base can be the direction of training (specialty), subject, theme and educational network. These types of elements and their mutual relations form the knowledge base structure, which is hierarchical. So, at the first level of the knowledge base there are specialties, at the second - the subjects to be learned, and at the last level there are sections of educational subjects, which, in turn, contain adaptive semantic models of educational material.

The system has a database, which consists of sixteen tables, the connection diagram of which is shown in Fig. 4.

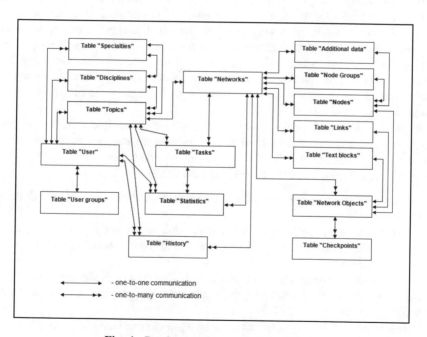

Fig. 4. Database tables connection scheme.

As you can see from the diagram, the main table of the database is the table "Networks". It brings together in a single structure all the other tables in the database.

The above diagram shows the complexity of the database structure. Such a database structure not only allows to automate the process of teaching students, but also to take into account the model of the student.

We have also developed and implemented a system of control of students' knowledge, based on adaptive semantic models and a network of request for educational information, providing: the appropriate sequence of presenting the student control tasks, the ability to use the activity approach to the process of control of knowledge and more complex tasks in the subject area, including in problem situations.

In addition, the project developed guidelines for the development and use of adaptive semantic models in various subject areas. The created intelligent system also allows you to save the user's history (date and time of entry into the system, the list of studied topics, received estimates, statistics of errors, etc.).

The above properties of the developed system are especially important in the organization of training in the electronic environment.

5   Conclusion

Thus, the use of intellectual information systems or their components in the field of education will provide access to subjects of the educational process to professionally-oriented knowledge bases, which are part of the architecture of IISEP. In addition, intellectual technologies in information systems for educational purposes are able to support the construction of a sequence of individual training courses, intellectual analysis of the answers of students, to provide interactive support in solving problems. Such systems, in particular, include: expert training systems; integrated intelligent systems based on integration of hypertext/hypermedia technologies and expert systems; systems based on hypertext and hypermedia technologies; intelligent multi-agent information systems.

In this article some problems of intellectualization of information systems of educational appointment and ways of their improvement were covered. Our research into the use of intelligent information systems in education continues. Within the framework of this project in the future it is planned to:

- Formation of modern information environment of the College using the developed intellectual information system.
- Development of the model in the form of a multilevel hierarchical semantic network corresponding to the structural and logical scheme of the curriculum of a particular area of training.
- Improvement of management of education on the basis of model representations of knowledge base of educational appointment, for the purpose of creation of electronic University.
- Development of an integrated intellectual system based on the synthesis of geographic and information-analytical systems.

References

1. Jain, L., Wu, X.: Advanced Information and Knowledge Processing. Springer, London (2009)
2. Wenger, E.: Artificial Intelligence and Tutoring Systems: Computational Approaches to the Communication of Knowledge. Morgan Kaufmann, Los Altos (1987)

3. Uskov, V.L., Uskov, A.V.: Web-based education: strategic issues for 2008-2015. Adv. Technol. Learn. **5**(1), 1–11 (2007)
4. Uskov, V.L., Uskov, A.V.: Modern technology-based education: instructor's profile. Adv. Technol. Learn **4**(3), 118–125 (2007)
5. Reichert, M., Oberhauser, R., Grambow, G.: Advances in Intelligent Process-Aware Information Systems, pp. 193–223. Springer (2017)
6. Vagramenko, Y.A., Yalamov, G.Y.: Collective educational activity of pupils in the network information and education environment. Pedagogical Informatics. No. 3, pp. 42–51 (2015)
7. Shikhnabieva, T., Beshenkov, S.: Intelligent system of training and control of knowledge, based on adaptive semantic models. In: Smart Education and e-Learning 2016, pp. 595–603. Springer International Publishing (2016)
8. Rod, M.: Implementing knowledge negotiation. In: Proceedings of the International Conference of Advanced Research Computer Education, Tokyo, 18–20 July 1990, pp. 321–328 (1990)
9. Szczerbicki, Edward, Graña, Manuel, Posada, Jorge, Toro, Carlos: Current research advances and implementations in smart knowledge-based systems: Part I. Cybern. Syst. **44**(2–3), 95–97 (2013)
10. Graña, M., Toro, C., Posada, J., Howlett, R.J., Jain, L.C. (eds.): Frontiers in Artificial Intelligence and Applications, Advances in Knowledge-Based and Intelligent Information and Engineering Systems, vol. 243 (2012). ISBN 978-1-61499-104-5
11. The ECAR Study of Undergraduate Students and Information Technology, EDUCASE Center for Applied Research, vol. 7, December 2006
12. Uskov, V.L., Uskov, A.V.: Modern technology-based education: instructor's profile. Adv. Technol. Learn. **4**(3), 118–125 (2007)
13. Vagramenko, J.A., Jalamov, G.J.: The Intellectualization of information systems to be included in the educational environment. Inf. Educ. Sci. **4**(32), S.3–S.11 (2016)
14. Danilyuk, S.G.: The analysis of uncertainty of problems of decision-making in intellectual problem-oriented educational systems. Scientific notes of IIO of Russian joint stock company, No. 50, pp. 49–69 (2013)
15. Rybina, G.V.: The intellectual training systems on the basis of the integrated expert systems: experience of development and use. The Information and measuring and operating systems, No. 10, pp. 4–16 (2011)
16. Shvetsov, A.N.: Construction metamethodology the multiagentnykh of intellectual systems. Information technologies in design and production, No. 1, pp. 28–33 (2010)
17. Shvetsov, A.N., Sibirtsev, E.V., Andrianov, I.A.: The computer training systems: multiagentny approach (An electronic resource). XII All-Russian meeting on problems of management VSPU-2014 22 Dec 2016. http://vspu2014.ipu.ru/proceedings/vspu2014.zip
18. Shikhnabiyeva, T.S., Brezhnev, A.V.: About one of options of development of the system of improvement of quality of management of education. Management of education: theory and practice, No. 3(27), pp. 50–57 (2017)
19. Shikhnabieva, T.S.: Training system and knowledge control based on adaptive semantic models. International scientific-practical conference "Development of national system of Informatization of education in health terms". The Information environment of education and science, No. 17 (2013). http://www.iiorao.ru/iio/pages/izdat/ison/publication/ ison_2013/num_17_2013/
20. Rastrigin, L.A., Erenshtein, M.K.: Adaptive learning with the learner's model, 160 p. Zinatne, Riga (1988)

Smart Interactive System for Learning of Tax Planning

Natalya V. Serdyukova[✉] and Kirill Potapov

Institute of Business Studies of the Russian Presidential Academy of National
Economy and Public Administration of the Russian Federation, Moscow, Russia
nat200612@yandex.ru

Abstract. This article describes principally new for the Russian practice
course, which serves for studying of taxation on the basis of a newly created
computer program – "Smart Presentation". It was utilized at first for the teaching
of subject "International tax planning" for 2 consecutive years (2016–2017) to
masters groups of the second year of studies at the Institute of Business Studies
of the Russian Presidential Academy of National Economy and Public
Administration of the Russian Federation. The innovation is that the program
modules a situation, in which a virtual enterprise is created and students through
the management of the activities of this enterprise receive certain tax consequences. In case of achievement of negative results of the business activity
student can return to the previous business case proposed. Implementation of the
program into learning process brought quiet positive results. The studying
efficiency indicator increased even up to 100% (in some groups). This experience represents first attempt of implementation of e-learning technologies for the
teaching economic subject - taxation. Creation of an educational game on the
basis of "Smart Presentation" now is under development. In frames of this game
teams of students can manage virtual companies, making decisions without the
possibility to change given answer and at the end the results of the teams will be
compared on a competitive basis. Also "Smart Presentation" can help any teacher, even without any skills in programming, to make interactive presentations
for students on any subject. Description of the technical decisions used is
highlighted in the article.

Keywords: E-learning technologies · International tax planning
Interactive presentation

1 Introduction

The experience, which is being described in this article, began 3 years ago (in 2015),
when one of the co-authors started teaching subject "International tax planning" to
masters groups of the second year of studies at the Institute of Business Studies of the
Russian Presidential Academy of National Economy and Public Administration of the
Russian Federation (hereinafter – the IBS).

The mission of the IBS is to educate and form the new generation of socially
responsible entrepreneurial and managerial elite of Russia and to advance Russian
business education.

In accordance with the mission the IBS provides its students with possibilities of studying of different subjects as "International tax planning" theory in which is closely interrelated with practice.

2 Background

As the outcome of studies of the subject IBS tends to receive a capable potential employee, who combines theoretical knowledge with practical experience of a tax consultant and will be able to cope with professional tasks in a frames of multinational holding company.

The Institute of Business Studies of the Russian Presidential Academy of National Economy and Public Administration of the Russian Federation implements this combination already in cooperation with such large and well-known holding companies as the PJSC "Sberbank" and the PJSC "LUKOIL". It has a joint master's program with the PJSC "Sberbank" and Memorandum on cooperation with the PJSC "LUKOIL". The cooperation with the PJSC "LUKOIL" consists in provision of benefits for the employees of the company and their children, joining the educational programs of the IBS on the monetary bases.

During the start of the studying process the following difficulties were highlighted:

1. The main task of the course was provision of students, who mainly previously haven't any practical or theoretical experience in taxation, with a knowledge of quiet high quality of the mechanisms and instruments of the international tax planning useful in the legitimate tax optimization processes of an enterprises.
   Quiet complicated massive of knowledge had to be provided to the recipients in quiet short terms (6 classes with duration of each for 3 astronomical hours).
2. During these 3 years the work was conducted with 6 different groups. Most of the groups consisted of the Russian students as well as of foreign (Spanish, German, French and Bulgarian) in proportion 50 to 50%.
   This obstacle highlighted the 2nd difficulty – difference in the educational systems of the countries of origin of students and heterogeneous in basic knowledge received by them.
3. The 3rd key difficulty was connected with organizing of uninterrupted system of control over the quality of the knowledge received during the whole course, consisting from techniques of verification of the interim and final results.
4. As well there was a problem with the search of qualitative source material to be used by students at home. "International tax planning" is a specific subject on which there are no too many updated textbooks, describing the subject generally.
5. Also must mention that due to the course students had to obtain basic skills in application of tax planning methods (such as: making the choice of jurisdiction for incorporation of a parent or subsidiary company, connected with the business goals of the group of holding companies, learn to calculate potential compulsory tax burden on the group of companies due to the choice made, get acquainted with mechanisms of avoidance of double taxation of the sole income (dividends, interest

payments, royalties and capital gains) in cross-border transactions, introduced BEPS measures, performance of the transfer pricing analyses, APAs, tax rulings, horizontal tax monitoring and fiscal unity.

The educational program provided was designed for the obtainment by students of concrete useful for business skills.

These programs form a new trend in development of educational process closely connected with business practice and in totally digital era are inseparably connected with the shift pro the vast use of smart educational technologies in a classroom.

As it was accentuated in the work [1] huge opportunities in terms of learning technologies, learning processes and strategies, corporate training, user's personal productivity and efficiency and faster and better educational services are provided due to the vast development of smart technologies.

In our work we mainly touch upon e-learning technologies in connection with the studying of such complicated economic subject as taxation.

We tried to make sampling of e-learning programs already used in the studying of economic subjects and found such programs implemented in Warsaw School of Economics, for example [2]. But mostly such courses are represented in a form of online lectures supplemented by additional materials such as tasks, tests and exercises. We haven't found examples of analogical educational quests used for the studying of taxation or international taxation. Our educational quest "Smart interactive program of tax planning learning" represents a game in course of which student creates and manage the activity of the virtual enterprise and deals with the tax consequences of the decisions taken. Also student in a course of the quest can correct the mistakes made due to the possibility to return to the previous stage of decision-making. All this will be observed in more details below.

3   Main Results

All these caused a need of search of optimal methods of teaching process to be used and required creation of optimal learning program on the subject, which I made an attempt to work out through these three years.

The program was divided into blocks, the control over each led to different results. Each block took about 1/3 of the whole time of the course.

The listed blocks were introduced:

1. The 1st block concluded initial theoretical base, which served as a basis for the subsequent solution of the further practical cases. It included verbal lectures and accompanying presentations in the Power Point format (no innovation here). Supervision over the process of studies shows that the effectiveness of this method represented in the assimilability of knowledge is at the medium level. All conclusions will be illustrated below with the results of the work of the groups. Recommendations for the designing of presentations can be found in [3–5].

2. The 2nd block consisted from the use of developed and implemented into educational process computer program which received name "Smart Presentation". This program allows teacher to create educational games and tests for a students

based on presentations designed in PowerPoint. A package of such presentations used within the course was called "Smart interactive program of tax planning learning" – innovation and the main scope of this research (see Fig. 1). It consists of interactive presentations due to the use of which a student deals with interactive situations connected with the business activity of a holding group, uses gained at the 1st part of the classes theoretical knowledge and receives different implications due to decisions made.

Fig. 1. Presentation package, starting page

The program represents different cases of the economic activity of the holding group, connected with their cross-border activity:

- Selection of proper jurisdiction for incorporation of a new subsidiary for this or that purpose;
- Performance of dividend payments by subsidiary companies;
- Performance of interest payments;
- Preparation of transfer pricing documentation with the use of the method proper for concrete transaction made;
- Formation of a fiscal unity;
- etc.

and legal consequences of each step due to the decision chosen by the student (see Fig. 2).

On each step student has a possibility to reconsider previously taken decision and to return the backward position. So this introduces 2 elements in the learning process:

- entertaining and fascinating component leading to better comprehensibility of knowledge;
- and continual recurrence of the material.

If a student took a decision, which led to negative consequences in the business activity of an entity, he becomes acquainted with them. Can return to the choice option and make decision again and thus step by step move further (see Fig. 3).

It's also important to stress that the contents of the sections can be easily changed, assuming needs. Also it is easy to create tests to monitor the progress of students.

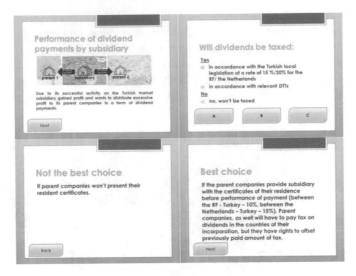

Fig. 2. Example of the part of the game, created with the use of Smart Presentation

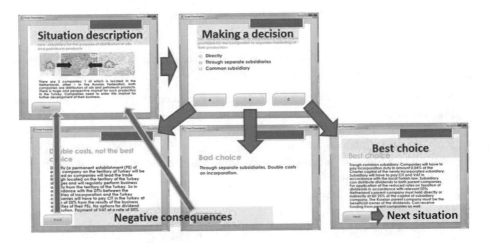

Fig. 3. Smart learning of tax planning in work

Technologically "Smart Presentation" consists of 2 modules:

- **design module**, which is integrated into Microsoft PowerPoint and allows to add UI components into presentation (for example, buttons) and assign the desired behavior to them (for example, "go to specified slide on mouse click");
- **execution module**, which is a standalone application and demonstrates the presentations created in the design module.

The design module provides a special toolbar, integrated into PowerPoint. This solution gives the well-known interface for the teacher, so he can create presentations as usual, adding interactivity with the help of controls. If the presentation is usually demonstrated slide-by-slide, then using "Smart Presentation" it is easy to turn the process into a test that counts the number of correct answers, and into an educational game, in which it is suggested to take a decision, and in case of an incorrect answer explained what the matter was. After that the user is returned to the question (the previous slide) to take on a new attempt.

In fact, "Smart Presentation" provides possibilities for programming presentation, but contrary to the integrated programming language VBA doesn't require special competencies from the user. Graphic interface of the module allows not only tuning of the outward of the UI components and prescription to them simple actions, but also creation of expression trees for calculation in more complicated cases (see Fig. 4). Of course, user's parameters are supported as well (for example, correct answers counter). All interactive elements of the presentation are constantly validated, which allows correcting errors in time.

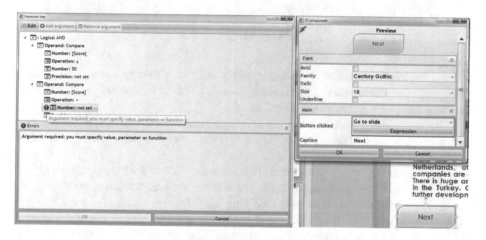

Fig. 4. Expression tree and button settings

The design module is implemented as a PowerPoint add-in [6], and the execution module as a standalone application. Both modules are written on C# language on the basis of platform .NET (version 4.0 – for the design module and version 3.5 – for the execution module), so the program could run on Windows 7 «out of the box» and use WPF for rendering user's interface.

Since presentations can be made in different styles it was important to provide the user with a possibility of customization of the outward of the UI components in accordance with the style of presentation. Flexibility and simplicity are ensured due to the use of XAML-based resource dictionaries (a lot of themes for WPF components can be found at the Internet and "Smart Presentation" allows to use them simply by coping into settings folder; see Fig. 5). WPF best practices and recommendations can be found in [7].

The design module builds zip package with all information for its demonstration after creation of the presentation, addition of the UI components into desired places and prescription of the necessary logic. This package isn't connected with the PowerPoint.

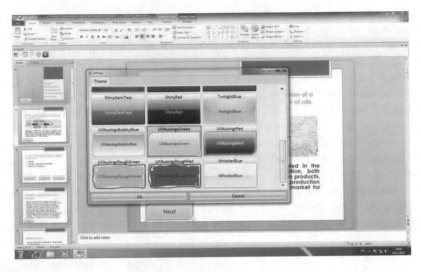

Fig. 5. Selection of the UI theme

In fact, the module renders presentation slides without UI components, complements them with the theme used (resource dictionary on XAML), serialized data on the interface and logic in xml document format.

The execution module opens zip package and reproduces the presentation. It doesn't require installation and can be deployed on any computer with .NET Framework 3.5 or later.

However, there are many programs that allow creation of training courses and tests. But Microsoft PowerPoint usually serves as habitual tool for a teacher, that's why in a course of our research we have observed plug-ins for this program. The most powerful from such programs is iSpring Suite [8]. Despite the fact that it is quite expensive ($720), the capabilities of this package are wider than any alternative (for example, free Office Mix [9]).

The iSpring Suite allows teacher to create full-fledged studying courses with the use of speech and video, but the possibilities for interactive presentations are small. You can insert into presentation some complex objects like "book" or "timeline" and create quizzes. Any quiz is a sequence of questions on the predefined forms, so you can't integrate the quiz question into slide, as well as turn the quiz into a full-fledged gaming application with its own logic.

In comparison with our plug-in there is no tool in iSpring Suite allowing creation of a set of screens with conditional transitions between them (as in text quests). Even the opportunity to describe why an answer was incorrect is limited by the design of the quiz (only a textual explanation). Of course, a hyperlink to the other slide of the PowerPoint presentation can be created, but addition even of the simplest condition requires programming skills. Therefore, the capabilities of iSpring Suite and similar plug-ins are not enough to create a full-fledged gaming applications based on the Power Point presentation.

However, iSpring Suite and some other plug-ins (for example, PPT to HTML5 [10]) can save presentations in a portable format (in this case – html5 with JavaScript). This allows to transfer the presentation to a computer without PowerPoint and run it immediately (so the other options offered by iSpring Suite are less convenient because they require installation of Adobe Flash). Since teachers quite often deal with different computers, it is better than the interactive presentation is independent from any modules that aren't included into operation system set.

3. The 3rd block – control of quality of the received on the subject knowledge.

Schematically the whole studying process can be depicted as follows – see Fig. 6.

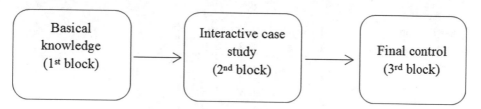

Fig. 6. Scheme of studying process

At first (in 2015), when the program was under development the studying process contained: 1st block, traditional case study without developed computer program and 3 stages of control (2 – interim and final).

From 2016 till 2017 process as well contained this newly introduced and worked out computer program. So we have the IBS internal data, gathered on the final results of the students of 6 masters groups (2 groups for each year), - for comparison and calculation of the effectiveness of the introduced innovation. This is reflected in the Table 1.

It's important to mention that in general points of the scoring scale correspond to the following (in accordance with the Provision on assessment of academic results of students of the RANEPA): 85–100 – "excellent"; 70–84 – "good"; 55–69 – "satisfactory"; 54 and less – "unsatisfactory".

These data shows an improvement of the indicator of the efficiency of the studying process in 2016–2017 (the years, in which innovative "Smart Program of tax planning learning" was additionally used in the process) in comparison with 2015. Thus, in 2015 the number of students who received marks "excellent" and "good" comprised in the 1st group – 85.2% and 57.1% in the 2nd. In 2016 academic level reached in the 1st group – 87.5% and in the 2nd – 77.8%. In 2017 respectively: 71.4% - in the 1st group and 100% - in the 2nd.

Table 1. Changes in the efficiency of studying process

Results in accordance with scoring scale	Year of studies	Groups		Final rank of the groups on the basis of efficiency indicator of the studying process[a]
		1	2	
Excellent	2015	10	0	3
Good		13	6	
Satisfactory		2	0	
Unsatisfactory		1[b]	1[b]	
Number of students		27	7	
Excellent	2016	2	4	2
Good		5	3	
Satisfactory		1	1	
Unsatisfactory		0	1[b]	
Number of students		8	9	
Excellent	2017	1	4	1
Good		4	5	
Satisfactory		1	0	
Unsatisfactory		1[c]	0	
Number of students		7	9	

[a]Results in accordance with scoring scale
[b]Due to the pass of the final test
[c]Hasn't attended the course

4   Conclusions and Next Steps

1. At the moment, "Smart Presentation" allows teacher to create interactive presentations and tests for knowledge control. In the future, program can be used to teach programming basics if list of supported UI components and actions will be expanded.

   In addition, it is planned to introduce a competitive element in the course. For this purpose an educational game is developed on the basis of "Smart Presentation". In frames of this game teams of students can manage virtual companies, making decisions without the possibility to change answer and at the end the team results are compared.

2. But, of course, "Smart Program of tax planning learning" is only one of the multiple factors in the whole learning process such as: general level of student's preparation, diligence, attendance of classes and self-preparation.

   It helps student to become involved in the tax planning process as a tax specialist of basic level and gain quiet unique experience as a practician and supports preparation of specialists in taxation with skills necessary for holding multinational companies.

3. Program "Smart Presentation" itself can be used for preparation of analogical quests on other subjects as well.

References

1. Uskov, V.L., Bakken, J.P., Pandey, A.: The ontology of next generation smart classrooms. In: Uskov, V.L., Howlett, R., Jain, L. (eds.) Smart Education and Smart e-Learning. Smart Innovation, Systems and Technologies, vol. 41. Springer, Cham (2015)
2. http://www.sgh.waw.pl/en/Pages/default.aspx
3. International Bureau of Fiscal Documentation. http://www.ibfd.com. Accessed 08 Feb 2018
4. http://eng-ibda.ranepa.ru/
5. Kapterev, A.: Presentation Secrets: Do What You Never Thought Possible with Your Presentations. Wiley, New Jersey (2011)
6. Office and SharePoint Development in Visual Studio. http://msdn.microsoft.com/en-us/library/d2tx7z6d.aspx. Accessed 08 Feb 2018
7. Nathan, A.: Windows Presentation Foundation Unleashed. SAMS Publishing, Indiana (2013)
8. iSpring homepage. http://www.ispringsolutions.com. Accessed 08 Feb 2018
9. Office Mix homepage. http://mix.office.com/en-us/Home. Accessed 08 Feb 2018
10. PPT to HTML5 homepage. http://www.ppt-to-html5.com. Accessed 08 Feb 2018

Smart Education Analytics: Quality Control of System Links

Natalia A. Serdyukova[1(✉)], Vladimir I. Serdyukov[2],
and Vladimir A. Slepov[1]

[1] Plekhanov Russian University of Economics, Moscow, Russia
nsns25@yandex.ru, vlalslepov@yandex.ru
[2] Institute of Management of Education of Russian Academy of Education,
Moscow State Technical University n.a. N.E. Bauman, Moscow, Russia
wis24@yandex.ru

Abstract. One of the most important questions in the development of the work of the system of Smart Education is the problem of quality control of its functioning, and therefore the questions of system connections. Links in Smart Education are characterized by different types of assessments, such as exam grades, matriculation certificate obtained at the end of school, ranking assessments, etc. So, to analyze links one needs to use method that can combine qualitative and quantitative characteristics. We propose to study connections of Smart Education System using the method of algebraic formalization. This method makes it possible to describe the hierarchy of system's structural links, to set off the levels of the system's links, and to prove theorem on the description of the system's links of each finite level. Different types of system's links are singled out. The notion of a factor - fractal system on the links of finite levels are defined. The classification of the system's links based on the number of synergetic effects of the group of the system's links of a finite level is constructed. There are several practice applications of this method in the paper. Thus, the justification of Granovetter hypothesis on the connections in social systems has been obtained. The next example relates to Smart Education. The structure of ranking systems for evaluating the effectiveness of the functioning of Smart University is proposed based on the formalization of the process of decomposition of system. Tensor efficiency index is introduced.

Keywords: Smart system · Algebraic formalization · Tensor estimate

1 Introduction. Smart Education System's Links Research

The concept of Smart Education differs from other educational concepts in the abundance of intrasystem connections of different levels. There are a lot of works on the principles of the systems approach, [1], on the concepts of equilibrium and synergetic approaches, [2, 3], on the key positions of synergetics, expressing its main content, [4], on algebra, [5], which show the importance the system's links. Nevertheless, there are no precise methods describing system's links. Therefore, we shall firstly consider the problem of describing system's links. First, let's consider the question of describing the hierarchy of system's structural links. The main construction runs as follows.

© Springer International Publishing AG, part of Springer Nature 2019
V. L. Uskov et al. (Eds.): KES SEEL-18 2018, SIST 99, pp. 104–113, 2019.
https://doi.org/10.1007/978-3-319-92363-5_10

Let's designate: S – the designation of a system; $S_0(S)$ - elements of a system; $C_0(S)$ – links between elements of a system, that is between elements of the set $S_0(S)$. Since links are relationships between elements of a system, for binary links of the first level of the system S, i.e. for connections between two elements, we have: for a binary link $\propto$ between elements of a system: $\propto \in S_0(S) \times S_0(S)$. For n-ary link of the first level of the system S we have: $\propto \in S_0(S) \times S_0(S) \times \ldots \times S_0(S)$, where the number of factors on the right-hand side is n. Let $P_1(S) = \{S_0(S); C_0(S)\}$ be a two-element set whose elements are the set of elements of the system S and the set of links $C_0(S)$ of the elements $S_0(S)$ of the system S. Let $\boldsymbol{P}_1(\boldsymbol{S}) = \langle P_1(S), \propto_1 \rangle$, where $\propto_1$ is a partial order on the set $C_0(S)$, that is $\propto_1$ structures the set of links $C_0(S)$ of the elements of the set $S_0(S)$. We can assume so, because one can go from a partially ordered set to a lattice in accordance with [5]. We call $\boldsymbol{P}_1(\boldsymbol{S}) = \langle P_1(S), \propto_1 \rangle$ the structure of the constraints of the first level of the system S. Let us construct now the second-level links of a system. Let $P_2(S) = \{P_1(S); C_1(S)\}$ be a two-element set whose elements are sets $P_1(S)$ and $C_1(S)$, $C_2(S)$ are links between elements of the set $C_1(S)$, that is $C_2(S)$ structures links of the level 1 of the system S. Then $C_2(S)$ are the links of the level 2 of the system S. For the second –level links of the system S we have: - for binary relation $\propto_2$ we have: $\propto_2 \in C_1(S) \times C_1(S)$, - for n –ary relation $\propto_2$ of the second level of the system S we have: $\propto_2 \in C_1(S) \times C_1(S) \times \ldots \times C_1(S)$, where the number of factors on the right-hand side is n. Now suppose that at the step n links' structure $\boldsymbol{P}_n(\boldsymbol{S}) = \langle P_n(S), \propto_n \rangle$ of the system S of the level n has already been constructed. Here $P_n(S) = \{P_{n-1}(S); C_{n-1}(S)\}$ is a two-elements set, whose elements are sets $P_{n-1}(S)$ and $C_{n-1}(S)$, and $\propto_n$ is a binary relation of a partial order on the set $C_{n-1}(S)$, that is $\propto_n$ structures the set of links $C_{n-1}(S)$ of the set $P_{n-1}(S)$, $C_{n-1}(S)$ are links of the level $n - 1$ of the system S. Elements of the set $C_n(S)$ are links of elements of the set $C_{n-1}(S)$; $C_n(S)$ are links of the level n of the system S. Then $P_{n+1}(S) = \{P_n(S); C_n(S)\}$ is a two-element set whose elements are sets $P_n(S)$ и $C_n(S)$, $C_{n+1}(S)$ are links of elements of the set $C_n(S)$, and $\propto_{n+1}$ is a relation of a partial order on the set $C_{n+1}(S)$, that is $\propto_{n+1}$ structures links of the level n of the system S. $C_{n+1}(S)$ are links of the level $n + 1$ of the system S. So, we obtain that links of the level k of the system S are defined for any natural number by the axiom of mathematical induction.

Definition 1.1. The system $\boldsymbol{P}_1(\boldsymbol{S}) = \langle P_1(S), \propto_1 \rangle$ is called the first level system dual to the system S.

This construction can be illustrated with the help of the following scheme.

Let S be any system, $S_1(S) = \{a, b, \ldots, c, \ldots\}$ be the set of elements of the system S, $C_1(S) = \{v, w, \ldots, z, \ldots\}$ be the set of links between elements of the system S. Links of the first level of the system can be illustrated with the help of the graph of the system:

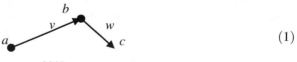

$$(1)$$

The edges of the graph (1) are elements of the set $C_1(S)$, representing links of elements of $S_1(S)$ of the system S. At the second level we consider links $C_2(S)$ between the elements of the set $C_1(S)$, which can also be represented by a graph:

$$(2)$$

This process is then continued by induction.

The following practice example from the field of Smart Systems justifies the consideration of links of different levels in systems and having been used in e-learning. Interdisciplinary links reveal the links between the relationships of properties in different subject areas, [6–11].

Theorem 1.1 About the Description of the System's Links. Links of the level no more than n of the system S, where n is a natural number, are determined by no more than two combinations of connections of the level no more than n of the system S.

The approach of allocating connections of various levels in a system is justified from the point of view of practice, since: - the system approach is applied in the case when it is a question of the description of complex systems, i.e. when the behavior of the system cannot be described with the help of a single mathematical model, - there are numerous examples showing the existence of such links in different systems, and, in particular, in Smart Systems. One needs to use such technic when examining economics of smart education, smart assessment and testing, smart e-learner modeling, smart e-learning management, and so on.

2 Types of System Connections. Operations Over System Links

The theory of Malt'sev's binary relations, [5], allows to build a classification of systems' binary links of any finite level and introduce operations on system's links. We can define reflexive binary relation of the first level of the system, symmetric binary relation of the first level of the system, antisymmetric binary relation of the first level of the system by the analogues with [5]. After that we can define a binary relation of equivalence on the set $S_0(S)$, that is $\propto$ is a reflexive, symmetric and transitive binary relation on the set $S_0(S)$. So, it is possible to eliminate the link α of the first level of the initial system S by considering the factor-system $S_0(S)/\propto$ with respect to the equivalence relation $\propto$ of the first level of the system S. The same operation can be done for any finite level of the system. Since we consider the connections of the system S as a relation between the elements of some set $C_n(S)$, this approach allows us to use the main results of AI. Malts'ev, [5], which concern operations over relations. Hence, we obtain: 1. The set of all binary links of a fixed level (realizable and unrealizable by the system under consideration) forms a Boolean algebra with respect to operations of union, intersection and taking complement of binary links. One can also speak of operations of inversion and multiplication of links in addition to operations $\cup, \cap, '$ over binary links of a fixed level. 2. Let the relation $\propto$ be defined on a pair of sets A, B and the relation β is defined on a pair of sets B, C. The product $\propto \cdot \beta$ of relations $\propto, \beta$ is called a relation defined on a pair of sets A, C such that $a \propto \cdot \beta c$ is true if and only if there exists an element x in the set B, such that $a \propto x$ and $x \beta c$ are true. If the relation $\propto$ is given on a pair of sets A, B, then the inverse relation or the inversion of a relation $\propto$ is called a relation $\propto^{-1}$, which is defined on a pair of sets B, A, consisting of all pairs (b, a), for which $(a, b) \in \alpha$.

So, we obtain the following assertion from 1 and 2:

Theorem 2.1. Let S be a system and $C_n(S)$ be a set of links of the level n of the system S. An algebra $C_n(S) = \langle C_n(S), \cdot, ^{-1}, e \rangle$, where $e = \{(a, a) | a \in C_n(S)\}$ is a group if $(\forall a, c \in C_n(S))(\exists x \in C_n(S))(a \propto x \& x \beta c)$ for every $\propto, \beta \in C_n(S)$.

This group is called a group of all links of the level n of a system S.

Different predicates P, given on the class of groups and closed under taking subgroups and factor- groups, determine the properties of the links of any fixed level of the system S, if these links themselves form a group. The number of synergetic effects of a closed associative system with feedback is determined by its number of final states, [12, 13]. So, if links of a system of some finite level k themselves form an associative closed system with a feedback of n_S^k elements, then the number of synergies of the system S is not exceeds the number of synergetic effects of the closed associative structure of the level k of the system S with a feedback if the following condition takes place. The number of pairwise nonisomorphic groups of order n_S^k is greater than the number of pairwise nonisomorphic groups of the order n_S. Here from one can obtain a classification of the links of a closed associative system with feedback in terms of the number of synergetic effects of the system S. Also, a classification of the links of a closed associative system with feedback in terms of and the number of synergetic effects of the closed associative structure of links of the level k of the system S with a feedback can be obtained in the same way. Undoubtedly, the construction of an exhaustive detailed classification of closed associative systems with a feedback even over the levels of the system's links is hardly possible now. Therefore, we consider a special case. Let us suppose that a group of factors G_S, which determines the system S, is finite. We also assume that for every positive integer n the set of links of the level n has a group structure, that is $C_n(S) = \langle C_n(S), \cdot, ^{-1}, e \rangle$ is a group. Here we are talking about a closed system of interacting factors that determine the system, and not a closed or open system in the classical sense of Theory of Systems. Let's designate: $g(n)$ is a number of pairwise nonisomorphic subgroups of a group of the order n. For example, if p is a prime number, then $g(p) = 1$. A partial classification can be obtained on specific examples which show how one can act in the general case within the framework of the assumptions made. Let's consider the following example, which is actual for practice. Let the group of factors G_S, which determine the system S, consists of p elements, where p is a prime number, $p > 2$. Then G_S defines the model of the system S without synergistic effects. Let's suppose, that the system S is arranged in such a way that each of its elements is connected with each other. So, the number of links of the first level $|C_1(S)|$ of the system S is not less than $p \times (p - 1)$. Besides it, let's assume that the links of each level k of the system S, where k is a natural number, also satisfy this condition. So, these links form a group, and each element of $C_k(S)$ is connected with each other. So, for each natural number n the following relation holds: $x_{n+1} = x_n(x_n - 1)$, where $x_n = |C_n(S)|$. Then the number of possible synergistic effects of the links of the system of each level k, where k is a natural number, is greater than 1. If $p = 2$, then the model G_S is the simplest model without synergistic effects at all levels of links. If, at some level of the links of the system S in the model G_S, we arrive at the situation where $|C_k(G_S)| = p = 2$, then the subsequent link levels of the system S do not contain synergistic effects according to the model G_S. These examples show that $p = 2$ plays a special role in the theory of systems, as well as in algebra in fields theory, as in the theory of numbers. Introduction of the concept of factors that determine the system, allows us to introduce the notion of factorial fractality in terms of links levels of the system.

Remark. It is possible to classify the finite models G_S of factors which determine the system S, in the case when for each positive integer n the links of the system of level n has a finite group structure, that is, $C_n(S) = \langle C_n(S), \cdot, ^{-1}, e \rangle$ is a group, since a complete description of finite groups has been obtained now, [see Atlas of finite groups].

Definition 2.1. The system S is called factor-fractal by levels i, j, if the group of links $G_i(S)$ of level i is isomorphic to the group of links $G_j(S)$ of level j of this system.

Such a fractality we encounter, for example, in biology when transferring properties from parents to offspring, in standards in smart education, in smart assessments and testing, and so on. The following algorithm of analysis and decomposition of the system by its links levels helps one in examining system's properties.

Algorithm 2. 1 of Analysis and Decomposition of the System by its Links Levels
Step 1. Single out elements $S_0(S)$ of the system S. Step 2. Single out links $C_0(S)$ between the system elements that is between the elements of the set $S_0(S)$. Step 3. Investigate the structure of first level links $P_1(S) = \langle P_1(S), \propto_1 \rangle$, where $\propto_1$ is a partial order relation on the set $C_0(S)$. Step 4. Continue this process by induction on levels of a system's links in accordance with the model of the hierarchy of structural links of a system....... Step 3n. Step 3n + 1. Step 3n + 2. Construct the structure of the links $P_n(S) = \langle P_n(S), \propto_n \rangle$ of the level n of the system S, where $\propto_n$ is partial order relation on the set $C_{n-1}(S)$. By the axiom of mathematical induction, this process can be continued to any set of links of the level n where n is a natural number.

At the same time, the direction of the algorithm for decomposing the system from the bottom to the top is from the lower level of communication to the upper one gives us an algorithm of a synthesis of a system.

3   Practice Example: The Social Relationships Strength

Let's consider connections in social systems. In 1940s Granovetter and Freeman explored social networks and introduced the notion of strong and weak social ties through the formal separation of two classes of interpersonal relationships by the frequency and duration of contacts. An example of strong ties are links with relatives and friends, the weak are the links between the neighbors, acquaintances, acquaintances of friends, formal contacts at work. Granovetter hypothesized that within social networks, weak connections are more important than strong ones.

Let's introduce the following definitions.

Definition 3.1. Let S be a system and G_S be a group of factors that determined the system S. The measure $PC(G_S)$ of the system S links strength is the number of possible different synergetic effects of the system S. So, $PC(G_S)$ is equal to the number of possible different final states of the system S, which are calculated by the model G_S, or, which is the same, to the number of pairwise nonisomorphic groups of order $|G_S|$.

Than $PC(G_S)$ index is larger, that the system S links calculated on the model G_S is weaker. Than $PC(G_S)$ index is smaller, that the system S links calculated on the model G_S is stronger. The system links indicator $PC(G_S)$ is a relative one and it depends on the choice of the model G_S of factors that determined the system S.

Definition 3.2. Let S be a system and G_S be a group of factors that determined the system S. Let $\emptyset \neq M \subseteq G_S$. The measure $PC(M)$ of the set M links strength is the number of possible different synergetic effects of the system $G_S \backslash M$, where $G_S \backslash M$ is a subgroup of the group G_S, generated by the set $G_S \backslash M$.

Than $PC(M)$ index is larger, that the set M links calculated on the model G_S is weaker. Than $PC(M)$ index is smaller, that the set M links calculated on the model G_S is stronger. The system links indicator $PC(M)$ is a relative one and it depends on the choice of the model G_S of factors that determined the system S.

The links strength indicators introduced in Definitions 3.1 and 3.2 explain Granovetter's theory.

Now let's state on several ways of formalizing the links of the system. The first way to construct a formalization of the system's links runs as follows. The visual representation of the connections of the system uses graph theory. We have constructed on this basis a group of the system's links that uses the Cayley graph of the group of factors G_S, determining the system S and the construction of the free product, [12]. The second way runs as follows. Let the system link connects some elements a, b of the system and we are examining the model of factors which determine the system S. Let this model be an algebraic system $A_S = \langle A_S, \Omega \rangle$ of the signature Ω. The system's links should preserve but not destroy the internal structure of the system. So, it is natural to consider homomorphisms of the system A_S into itself, that is maps of the set A_S into itself, preserving operations and predicates from Ω, as the system's links. Hence from we obtain several ways to study duality in smart systems theory. The first way of constructing duality for the theory of smart systems uses models of factors that determine the system, and these models of factors are algebraic systems $A_S = \langle A_S, \Omega \rangle$ of some signature Ω. Further the classical theory of duality from category theory is used in this method. The second method was proposed by us for the case when the model of factors is a group of factors G_S. Here we can consider the following two cases.

The first case. The group of factors which determine the system S is finite, and $|G_S| = n$. It is well known that in this case the group G_S can be embedded in the symmetric group of all permutations S_n of degree n. The second case does not use restrictions on the number of elements of the group G_S. In the second method, we propose to embed G_S in its holomorph $HolG_S$. First let's consider the case where G_S is a finite abelian group. The holomorph $HolG_S$ of the group G_S is a semidirect extension of the group G_S with the help of its group of automorphisms $Aut(G_S)$. In general, holomorph of a group is the concept of Group Theory that arose in connection with the solution of the following problem: is it possible to include any given group G as a normal subgroup in some other group so that all automorphisms of G are consequences of inner automorphisms of this larger group? Herewith the automorphisms $Aut(G_S)$ of the group of factors of the system S are in fact links of the system S with special properties: (1) the one-to-one correspondence between factors that determine the system, (2) the preservation by the link of the composition operation of the factors which determine the system. In this connection, a special role here belongs to perfect groups, that is such groups G which are isomorphic to the group of its automorphisms $Aut(G)$. For example, $G \cong S_n$, where $n \neq 2, 6$. We have $HolG/G \cong AutG \cong G$ for a perfect group G. The next theorem explains the role of duality in the theory of strong and weak links of the system.

Theorem 3.1. Let S. be a system and $G_S \cong V$, where V is an additive group of the Euclidean vector space of the dimension n, be a group of factors which determine the system S. Then the powers of links of the system S and the system S' dual to S and defined by the group of factors V', where V' is an additive group of the vector space V' which is conjugate or dual to the vector space V, are the same.

4 Formalization of Smart System's Efficiency

One of the key concepts of Operations Research is the concept of efficiency. To compare the states of the system, the efficiency criteria are used, according to which the comparison takes place. The criterion is a rule, according to which the possible states of the controlled system are compared. Formalization of the criterion of effectiveness is called an indicator of effectiveness.

Definition 4.1. An operation aimed at achieving state A is a system of actions which goal is to achieve state A.

One of the first representatives of management theorists G. Emerson, singled out the relationship between efficiency and functionality. This relationship allows us to determine the effectiveness as a utility function.

Definition 4.2. The efficiency function $u : X \to R$, defined on the set of all final states of the system S can be defined as a utility function with properties: for any $x, y \in X$, if $x \succsim y$, then $u(x) \geq u(y)$.

Let us give some examples of a possible definition of the efficiency function of the system S.

1. The efficiency of the system by the final states of the system. In the first example G_S is a group of factors which determined the system S, $X = \{G_i | i \in I\}$ is a set of all final states of the system S, S and G_S are finite, so X is finite. The relation $\succsim$ on the set X reflects preferences about the finite state of a system.

2. The efficiency of the system's links according to the levels of the system's connections. We set up $u(G_k(S)) \geq u(G_m(S)) \Leftrightarrow k \geq m$, where $G_k(S)$ and $G_m(S)$ are the links of the level k and m the system S respectively.
3. The efficiency of the system by links of the system of the same level. Let $a, b \in G_k(S)$ are links of the level k of the system S, $\propto_k$ is a relation of link preference at the level k. Then $u(a) \geq u(b) \Leftrightarrow a \propto_k b$.

5 Main Results. Tensor Estimate of the Efficiency of a Smart System

The tensor in mathematics is an object which is characterized by an array of indicators as follows. Let V be an $n-$ dimensional vector space, $n < \infty$. The dual space V' to the space V is defined as the space of linear functions from V into the field of real numbers R. If the dimension of the vector space V is equal to n, then the dimension of the dual vector space V' is also equal to n. Vectors from V are called contravariant, and vectors from V' are called covariant. If to each coordinate system in the n - dimensional Euclidean vector space V there corresponds a system of n^{p+q} numbers $a_{ij...}^{rs...}$, where the number of lower indices is equal to p, and the number of superscripts is equal to q, in such a way that when passing from one coordinate system to another, these numbers are transformed by the formula

$$a_{ij...}^{'rs...} = c_i^\alpha c_j^\beta \ldots b_\sigma^r b_\tau^s \ldots a_{\alpha\beta...}^{\sigma\tau...} \tag{5}$$

where $\left\| c_i^j \right\|$ is a matrix of the transition from one basis of the vector space V to another basis, and $\left\| b_i^j \right\|$ is a matrix which is transposed to the matrix inverse to the matrix $\left\| c_i^j \right\|$, then we say that a tensor is given which is p times covariant and q times contravariant. Particular cases of tensors are vectors and scalars. The foundations of the tensor calculus were established in classical works of K. Gauss, G. Grassmann, B. Riemann, E. Christoffel, G. Ricci-Curbastro, T. Levi-Civita. The tensor is an object of linear algebra that linearly transforms elements of one linear space into elements of another linear space. Besides it, the tensor is a mapping that allows one to concentrate huge information arrays. The task of constructing a numerical estimate of the effectiveness of the functioning of the system is extremely difficult from a mathematical point of view since its solution involves a quantitative assessment of the appearance of qualitative changes. Let's note that, in fact, tensor estimates are used in ranking systems for assessing the performance of universities. We shall construct a tensor estimate of the effectiveness of the functioning of the system as follows.

Definition 5.1 A tensor estimate of the effectiveness of the functioning of the system is a homomorphism of a group of factors G_S, determining the system S, into a group $GL(n, R)$ of linear homogeneous transformations of the vector space R^n, that is

$$f : G_S \rightarrow GL(n, R).$$

An assessment of efficiency of functioning of the innovation system can be carried out according to the following algorithm.

Algorithm 5.1 of a Complex Estimation of Efficiency of Functioning of the Smart System Step 1. The construction of quantitative indicators describing the functioning of the innovation system. Identifying and monitoring the compliance of the functioning of the innovation system with its purpose. Step 2. The construction of the tensor evaluation of the innovation system's functioning. Step 3. The identification and monitoring of all the links of the innovation system with the external super system and its subsystems that arise during the functioning of the innovation system. Step 4. The identification of internal and external attributive factors of the innovation system. Step 5. The construction of internal qualitative indicators of the functioning of the innovation system in the form of a graph and a group of internal attributive features of the innovation system. Step 6. The construction of external quality indicators of intra system connections of the innovation system in the form of a graph and a group of external attributive features of the innovation system. Step 7. The determination of the number of possible synergetic effects of the innovation system. Step 9. The construction a model an innovation system's functioning based on its internal and external attributive features. Step 10. The verification of the possibility of increasing the accuracy of the model by introducing additional factors into the model. Step 11. The correction the management actions to prevent the occurrence of undesirable synergistic effects.

6   Conclusion

The proposed smart analytics is important for practice, since it allows to describe the connections of any finite level of a complex system and means to manage the functioning of Smart Universities. To evaluate and manage the activities of a Smart University, we propose using ranking systems. At the same time, these ranking systems should differ significantly from the known ones. We propose the following scheme of construction of such systems.

1. The development of the blocks of indicators for Smart Universities. The development and the construction of optimization models for each group of indicators and, based on overall optimization model to plan reliably the development of the Smart Education System with the priority of sustainable development and continuity of Smart Scientific Schools.
2. The construction of the model and the plan of the optimal realization of Smart Education on the base of ranking system for Smart Universities.
3. The consideration of the establishment of a new Smart Ranking System which has the assessment of sustainability greater than that of well − known systems.

References

1. Mogilevsky, V.D.: Methodology of systems, Economics, Moscow (1999). (in Russian)
2. Prigozhyn, I., Stengers, I.: Order from chaos. A new dialogue between man and nature. Progress, Moscow (1986). (in Russian)
3. Ivanov, A.E.: Genesis of Synergetics, Electronic scientific and practical journal. Modern scientific research and innovations, 9 (2013), LNCS Homepage. http://web.snauka.ru/issues/2013/09/26327
4. Knyazeva, E.N.: Synergetic-30 years old. Interview with Haken, G., Questions of philosophy (3), 53–61 (2000). (in Russian)
5. Maltcev, A.I.: Algebraic systems. Nauka, Moscow (1970). (in Russian)
6. Serdyukova, N.A.: On Generalizations of Purities, Algebra & Logic, 30(4), 432–456 (1991)
7. Serdyukova, N.A.: Optimization of Tax System of Russia, parts I and II, Budget and Treasury Academy, Rostov State Economic University, in Russian (2002)
8. Serdyukova, N.A., Serdyukov, V.I.: The new scheme of a formalization of an expert system in teaching. ICEE/ICIT 2014 Proceedings, paper 032, Riga (2014)
9. Serdyukova, N.A., Serdyukov, V.I., Slepov, V.A.: Formalization of knowledge systems on the basis of system approach, SEEL2015, Smart Education and Smart e-Learning, Smart Innovation, Systems and Technologies, vol. 41, pp. 371–380. Springer, Cham (2015)
10. Serdyukova, N.A., Serdyukov, V.I.: Modeling, Simulations and Optimization Based on Algebraic Formalization of the System, 19th International Conference on Engineering Education July 20 – 24, 2015, Zagreb, Zadar (Croatia), ICEE2015 New Technologies and Innovation in Education for Global Business, Proceedings, pp. 576–582 (2015)
11. Uskov A.V., Serdyukova N.A., Serdyukov V.I., Byerly A., Heinemann C.: Optimal Design of IPSEC-Based Mobile Virtual Private Networks For Secure Transfer of Multimedia Data, Smart Innovation, Systems and Technologies, vol. 55, pp. 51–62. Springer, Cham (2016)
12. Serdyukova, N.A., Serdyukov, V.I., Slepov, V.A, Uskov V.L., Ilyin V.V.: A Formal Algebraic Approach to Modelling Smart University as an Efficient and Innovative System, SEEL2016, Smart Education and Smart e – Learning, Smart Innovation, Systems and Technologies, vol. 59, pp. 83–96. Springer, Cham (2016)
13. Serdyukova, N.A., Serdyukov, V.I., Uskov, A.V., Slepov, V.A, Heinemann, C.: Algebraic Formalization of Sustainability Ranking Systems for Evaluating University Activities: Theory and Practice, SEEL2017, Smart Education and Smart e – Learning, Smart Innovation, Systems and Technologies, vol. 75, pp. 459– 474. Springer, Cham (2017)

Smart Pedagogy

Learning Analytics Based Smart Pedagogy: Student Feedback

Vladimir L. Uskov[1]($\boxtimes$), Jeffrey P. Bakken[2], Lavanya Aluri[1],
Narmada Rayala[1], Maria Uskova[3], Karnika Sharma[1],
and Rama Rachakonda[1]

[1] Department of Computer Science and Information Systems, and InterLabs
Research Institute, Bradley University, Peoria, IL, USA
uskov@fsmail.bradley.edu
[2] The Graduate School, Bradley University, Peoria, IL, USA
jbakken@fsmail.bradley.edu
[3] Midstate College, Peoria, IL, USA

Abstract. Learning Analytics is a dynamic interdisciplinary field that encompasses educational sciences and state-of-the-art technology, methods and systems from various fields of computing such as data science, data visualization, software engineering, human-computer interaction, statistics, artificial intelligence, with various stakeholders, e.g. instructors, students, department and college administrators, practitioners, university top managers, computer scientists, IT experts, and software developers. Despite some current achievements and initial developments in Learning Analytics, we are still in a very early stage of development of sophisticated technologies and well-thought practices, tools and applications in this field as well as understanding the impact of Learning Analytics on (a) student learning and privacy, and (b) faculty instruction and autonomy. This paper presents the up-to-date research findings and outcomes of a multi-aspect project on Smart Learning Analytics at Bradley University (USA). It describes the obtained research outcomes about student perception and attitude to Learning Analytics on an academic course level and corresponding Learning Analytics-based pedagogy.

Keywords: Smart pedagogy · Learning analytics · Teaching strategy
Student feedback

1 Introduction and Literature Review

Learning Analytics (LA) technologies and their applications in academia are rapidly gaining popularity among academic institutions in the world.

Motivation of LA Development and Deployment. In accordance with the EDU-CAUSE (USA) report [1] this innovative information technology provides significant contributions to academic success, including (1) increased student retention (34% of 245 EDUCASE respondents highlighted this benefit), (2) improved student course-level performance (19%), (3) decreased time to degree (19%), (4) understanding the characteristics of student population (about 19%), (5) optimizing resources (about

© Springer International Publishing AG, part of Springer Nature 2019
V. L. Uskov et al. (Eds.): KES SEEL-18 2018, SIST 99, pp. 117–131, 2019.
https://doi.org/10.1007/978-3-319-92363-5_11

19%), (6) attracting more students (about 10%), (7) reaching a different or broader segment of students (about 10%), (8) creating greater transparency (about 10%), (9) improving faculty productivity (about 10%), and (10) improving administrative service quality (about 10%) [1].

According to the JISC (UK) report [2], "Learning analytics has the potential to transform the way we measure impact and outcomes in learning environments – enabling providers to develop new ways of achieving excellence in teaching and learning, and providing students with new information to make the best choices about their education. ... Learning analytics can provide students with an opportunity to take control of their own learning, give them a better idea of their current performance in real-time and help them to make informed choices about what to study" [2].

Hanover Research (USA) report [3] particularly reads, "As institutions of higher education explore ways to develop cohesive technology-enhanced learning strategies, one important, emerging element of those strategies is learning analytics that supports students' self-assessment of their academic progress. ... Although data analytics initiatives within higher education have traditionally focused on institution-wide applications, learning analytics is a growing area of interest for many HEIs. Existing learning analytics initiatives are commonly connected to student performance monitoring efforts, including initiatives to improve retention, increase course completion, and reduce time to degree completion".

Finally, "A recent report from the U.S. Department of Education makes the point that on the program and institutional level, learning analytics can play a role that is similar to that of already existing business intelligence departments and applications. Just as business intelligence may utilize demographic, behavioral and other information associated with a particular enterprise and its customers to inform decisions about marketing, service and strategy, learning analytics promises to do something similar in educational terms" [4].

Definitions of LA. Various authors of available publications define the goals of LA in different ways.

The Society for Learning Analytics Research defines LA [5] as: "... the measurement, collection, analysis and reporting of data about learners and their contexts, for purposes of understanding and optimizing learning and the environments in which it occurs".

In accordance with Siemens [6], "The broad goal of learning analytics is to apply the outcomes of analyzing data gathered by monitoring and measuring the learning process, as feedback to assist directing that same learning process. ...Six objectives are distinguished: predicting learner performance and modeling learners, suggesting relevant learning resources, increasing reflection and awareness, enhancing social learning environments, detecting undesirable learner behaviors, and detecting affects of learners."

Suchithra et al. [7] believe that "The main purpose of Learning Analytics is to improve the performance of learners. Also, the environment of learning in which the learner undergoes is enhanced which will ultimately result in a quality education. Learning Analytics helps educator/teacher to understand the students. Learning capabilities can be improved for the learners. ... The Learning Analytics aims at the curriculum design, predicting the students' performance, improving the teaching learning environment, decision support system for Higher Education Institutions, personalized approach to individual students, online and other learning modes including mobile, subject wise teaching and learning, subjects which has practical and evaluation process in the education system. ... Learning Analytics is about the collection, analysis of data about the learners. It is an emerging field in research which uses data analysis on every tier of educational system".

Borkar and Rajeswari presented the following approach in [8]: "Learning analytics approaches in general offer different kinds of computational support for tracking learner behavior, managing educational data, visualizing patterns, and providing rapid feedback to both educators and learners".

Additionally, Tempelaar et al. [9] argue that "The prime data source for most learning analytic applications is data generated by learner activities, such as learner participation in continuous, formative assessments. That information is frequently supplemented by background data retrieved from learning management systems and other concern systems, as for example accounts of prior education".

2 Our Previous Works in Learning Analytics and Smart Pedagogy Areas

The research fellows and associates at the InterLabs Research Institute at Bradley University (USA) are actively involved in a multi-aspect research project with a focus on the design of Smart Pedagogy (SmP) in general, and, particularly, on in-depth analysis of possible SmP components - innovative advanced technology-based teaching strategies. These teaching strategies include but are not limited (1) learning-by-doing, (2) flipped classroom, (3) gamification of learning, (4) collaborative learning, (5) learning analytics, (6) formative analytics, (7) adaptive teaching, (8) context-based learning, (9) "bring your own device" (BYOD) strategy, (10) personal enquiry based learning, (11) crossover learning, and (12) robotics-based learning. The obtained outcomes of our research, analysis and testing of several of the designated innovative teaching strategies undoubtedly prove that they strongly support various "smartness" levels and smart features of SmP, Smart Education (SmE) and Smart Classroom (SmC) concepts. These proposed concepts, developed conceptual models, unique approaches and an implementation of obtained research findings and outcomes are presented in our works [10–18].

We define Smart Pedagogy (SmP) as a set of instructor's teaching strategies, activities and judgements to (a) understand the student/learner profile (background, goals, skills and capabilities), and (b) provide optimal (smart) learning environment to help students to achieve their goals. This type of pedagogy is based on active use of "smartness" features of Smart Education (SmE), including (1) adaptation, (2) sensing, (3) inferring, (4) self-learning, (5) anticipation and (6) self-organization [10, 18–20].

The introduced "smartness" levels of SmE are described in [10, 16–18], the conceptual model of SmE is described in [18], and the learning environment of a new type – Smart Classroom (SmC) is described in [10–14, 16]. The analysis of implementation of several innovative pedagogies in SmC and obtained student feedback is presented in several of our works, specifically: (1) applications of *Learning-by-Doing, Flipped*

Classroom, and *Gamification of Learning* teaching strategies in SmC are described in [11], and (2) *Collaborative Learning* – in [18]. For example, Table 1 from [18] presents the identified correspondence between introduced "smartness" levels of SmE, and several analyzed innovative technology-based teaching strategies – components of SmP.

Table 1. SmE "smartness" levels and identified corresponding innovative teaching strategies (as described in our previous work [18])

SmE smartness levels	Innovative SmP component to support SmE/SmU/SmC smartness levels (examples)
Adaptation	*Gamification of Learning* allows a learner to adapt to attractive features of gamified courses such as extra points, badges, self-control and monitoring of academic performance as well comparing student performance with achievements by other students
Sensing	Mobile devices (as part of Bring-Your-Own-Device – *BYOD* – teaching/learning strategy) with built-in accelerometers, noise, light, humidity and temperature sensors, can be used as science toolkits to collect data and perform experiments in various courses
Inferring (logical reasoning)	*Learning Analytics (LA)* strategy provides students and faculty with just-in time raw data and processed information (in a form of data visualization and corresponding recommendations) about student academic performance
Self-learning	*Context-Based Learning (CBL)* as well as *Crowd Learning* help students/learners to improve learning as it enables people to learn from other users/students/learners. It also allows students to watch video lectures at home and discuss them in SmC *Collaborative Learning (CL)* helps to improve learning by enabling a more social form of study, with a group of students working together on various projects
Anticipation	LA helps student to know his/her academic performance (progress) in each course, and predict final score (grade) based on course statistics (or, course intelligence) from other students in the past
Self-organization	The main concept of CL is based on self-organized or organized virtual learning communities; it is actively used to improve communication, collaboration and team-working of students/learners

3   Project Goal and Objectives

The goal of the current phase of this research project is to analyze, implement and obtain student feedback regarding LA-based pedagogy on academic course level – the one that heavily uses various types of student-related data in an academic course.

The main functions of LA-based pedagogy (project objectives) on course level should include the following ones.

Instructors should be able to use LA to:

- discover student learning patterns or submission patterns;
- identify student profiles (for example, student motivation to complete optional assignments, time of submission of assignments – always one of the first submitters or always "last minute" submitter);
- explore student academic performance data;
- identify early indicators/signals for student potential success or failure, poor grades, potential drop-off, etc.;
- identify potential problems with specific students (for example, identify at-risk students – those who have poor academic performance at any given time of a course);
- predict student learning trajectory, final scores, final grades in a course;
- predict student analytical, technical, management and communication skills in a course;
- motivate, intervene, supervise, advise and assist students;
- thoroughly and in detail monitor learning processes and activities in a course;
- increase transparency of academic performance and awareness in a class;
- increase the effectiveness of learning process and/or learning environments;
- improve (adapt, change, modify) teaching strategies, educational resources and learning environments based on sensing and inferring "smartness" features of SmP and Smart Learning Analytics (SLA) system;
- provide detailed analysis in terms of descriptive, diagnostic, predictive and prescriptive analytics – major components of LA, and other functions.

Students should be able to use LA to:

- monitor their academic performance, learning activities, the entire learning process, and assess/compare his/her academic performance with performance of other students in a course (i.e. classmates);
- be informed about current level of his/her analytical, technical, communication, and management skills and comparison of his/her skills to skills of average current student in a course or average student in the past in this course;
- predict academic performance, identify new goals(s) in a course, and, as a result, improve his/her learning activities in a course; and
- identify a need for advising, tutoring, or coaching.

4   Learning Environment Used and Student Feedback Obtained

In 2016–2018 year one of the co-authors taught several Computer Science (CS) and Computer Information Systems (CIS) undergraduate and graduate courses to master various components of SmP, including (1) learning-by-doing, (2) flipped classrooms, (3) gamification of learning, (4) learning analytics, (5) formative analytics, and (6) collaborative learning. All those classes were taught in the Br160 Smart Classroom

[17, 18] – a highly technological and modern classroom at Bradley University. It is equipped with the advanced software and hardware systems and technology for SmP.

During the Spring-2018 semester, 23 undergraduate students in "Introduction to Software Engineering" and "Systems' Analysis and Design" undergraduate courses and 9 graduate students in "Software Project Management" course were surveyed on various components of LA-based pedagogy – pedagogy that was actively used in these classes. Students were asked multiple questions (part of them are presented in Tables 2 and 3 below) about LA-based pedagogy. We were interested, first of all, in the identification of percent of students who have a positive opinion about LA-based pedagogy; this is the reason that we did not use a traditional 5-point Likert scale in these surveys.

The following legend is used in Tables 2 and 3 to adequately represent students' answers:

- "1" – "I like it very much"; this type of answer contributes 3 points to average student rank of teaching strategy;
- "2" – "I somewhat like it"; this type of answer contributes 2 points to average student rank of teaching strategy;
- "3" – "Other opinions" (that, in general case, may include "I don't like it very much", "I somewhat don't like it", "I don't have clear opinion/preference about this teaching strategy"); this type of answer contributes 1 point to average student rank of teaching strategy;
- Av. UG – average score by all undergraduate students on 3-point scale, where 3.0 is associated with maximum positive "I like it very much" student opinion and "1" – "Other opinions"; additionally in this column we presented a total number (in %) of student answers of "1" and "2" types – together they represent positive student opinion about LA-based pedagogy;
- Av. GR – average score by all graduate students on 3-point scale; additionally in this column we presented a total number (in %) of student answers of "1" and "2" types – together they represent positive student opinion about LA-based pedagogy.

The obtained data regarding student feedback about LA-based pedagogy are presented in Tables 2 and 3 below.

Table 2. LA-based pedagogy: a summary of undergraduate (UG) students' feedback

Questions to survey students	1	2	3	Av. UG score and %
Descriptive Analytics of LA-based pedagogy ("What happened so far?")				
(1) Do you want to know your **academic performance for each learning assignment** in this course at any point of time in the semester?	81%	19%	0%	**2.81 (100%)**
(2) Do you want to know your **up-to-date overall academic performance** (total score or total number of points obtained so far) in this course at any point of time in the semester?	88%	4%	8%	**2.81 (92%)**

(continued)

Table 2. (*continued*)

Questions to survey students	1	2	3	Av. UG score and %
(3) Do you want to see or obtain data about **academic performance of other students** in this course (i.e. your classmates) under the condition that no student names will be associated with those data but just student ClassIDs? In this case, you will be able to compare your academic performance with the academic performance of your classmates	42%	38%	19%	**2.23** **(81%)**
(4) Do you want the instructor to **include data about your anonymized academic performance** (i.e., without using your actual name but only your ClassID) into the roster of academic performance of all students in this class? In this case, other students (your classmates) will be able to see your (without your actual name but anonymized ClassID) academic performance in this course	35%	38%	27%	**2.08** **(73%)**
(5) Do you want the instructor to **identify the up-to-date levels of students' analytical, technical, management, and communication skills** in this course?	62%	23%	15%	**2.46** **(85%)**
(6) Do you want the instructor to **clearly identify current at-risk students** (without actual names of students but only anonymized ClassIDs), i.e. students who currently have very low academic performance (usually, at the level of a "D" grade and below)?	50%	15%	35%	**2.15** **(65%)**
Diagnostic Analytics of LA-based pedagogy ("Why did it happen?")				
(7) Do you want to know why you have a **certain identified overall current score** (i.e. total obtained points) in this course so far?	54%	27%	19%	**2.35** **(81%)**
(8) Do you want to know why you have **certain identified current levels of analytical, technical, management and communications skills** in this course so far?	58%	31%	12%	**2.46** **(88%)**
(9) Are you willing to **stop by the instructor's office during office hours** to discuss the above-mentioned two topics of Diagnostic Analytics?	50%	27%	23%	**2.27** **(77%)**
(10) Did you try to stop by the instructor's office to **get help, get answers to your questions and/or improve your knowledge on specific topics** in this course?	15%	31%	54%	**1.62** **(46%)**

(*continued*)

Table 2. (*continued*)

Questions to survey students	1	2	3	Av. UG score and %
(11) Did you try to **complete optional homework learning assignments** (for extra points) in this course?	23%	38%	38%	**1.85 (62%)**
(12) Did you try to **submit/provide your solution for optional in-classroom assignments/** discussions (for potential extra points) in this course?	4%	38%	58%	**1.46 (42%)**
Predictive Analytics of LA-based pedagogy ("What happened so far?")				
(13) Do you want **the instructor today to predict your total final score** (and, as a result, your final grade) in this course?	54%	31%	15%	**2.38 (85%)**
(14) Do you want to see or **get data about predicted final scores and/or final grades of other students in this course** under the condition that no student names but anonymized ClassIDs only will be associated with those data? In this case, you will be able to compare your predicted academic performance with the academic performance of your classmates	54%	15%	31%	**2.23 (69%)**
(15) Do you want the instructor to **include predicted data about your anonymized final score and/or final grade** (i.e. without using your actual name but only ClassID) **into the roster of predicted academic performance of all students** in this class? In this case, all other students (i.e. your classmates) will be able to see predicted academic performance of all students (but without actual name of any student but student ClassID only)	38%	27%	35%	**2.04 (65%)**
(16) Do you want the **instructor today to predict levels of your analytical, technical, management, and communication skills** by the end of this course?	50%	27%	23%	**2.27 (77%)**
(17) Do you want the instructor today to **predict and show to all classmates** (without actual names of students but their ClassIDs only) **potential students who are predicted to be at-risk** by the end of this course (i.e. students with predicted low-level academic performance by the end of this course)?	27%	27%	46%	**1.81 (54%)**

(*continued*)

Table 2. (*continued*)

Questions to survey students	1	2	3	Av. UG score and %
(18) Do you want **to receive anonymous and confidential alerts via email if the Smart Learning Analytics (SLA) system identifies you as currently an at-risk student** or predicts you to be an at-risk student by the end of this course?	54%	19%	27%	**2.27 (73%)**
Prescriptive Analytics of LA-based pedagogy ("How can we make it happen?")				
(19) Do you want to be able to **set up a new goal** (for example, a new desired final grade or final score in this course by the end of semester) based on current outcomes of Descriptive and Diagnostic Analytics, and **simulate with the Smart Learning Analytics (SLA) system potential needed activities/achievements/scores for remained learning assignments**? (In other words, Are you comfortable getting the answer to the following question: Can I achieve that new goal by the end of this course or not?)	69%	19%	12%	**2.58 (88%)**
(20) Do you want **the instructor and/or tutor to provide additional guidance and/or recommendations on what you should do** to achieve your new academic goal (for example, new final grade in this course) and/or improve your academic performance accordingly (based on the calculated prediction) in this course?	69%	19%	12%	**2.58 (88%)**
(21) Are you willing to **spend more than recommended 3 hours per week of independent student work with a tutor** to improve your current academic performance in this course?	35%	27%	38%	**1.96 (62%)**
In general, what is your opinion about LA-based pedagogy	**48%**	**26%**	**26%**	**2.22 (74%)**

Table 3. LA-based pedagogy: a summary of graduate (GR) students' feedback

Questions to survey students	1	2	3	Av. GR score
Descriptive Analytics of LA-based pedagogy ("What happened so far?")				
(1) Do you want to know your **academic performance for each learning assignment** in this course at any point of time in the semester?	100%	0%	0%	**3.00 (100%)**

(*continued*)

Table 3. (*continued*)

Questions to survey students	1	2	3	Av. GR score
(2) Do you want to know your **up-to-date overall academic performance** (total score or total number of points obtained so far) in this course at any point of time in the semester?	100%	0%	0%	**3.00 (100%)**
(3) Do you want to see or obtain data about **academic performance of other students** in this course (i.e. your classmates) under the condition that no student names will be associated with those data but just student ClassIDs? In this case, you will be able to compare your academic performance with the academic performance of your classmates	43%	14%	43%	**2.00 (57%)**
(4) Do you want the instructor to **include data about your anonymized academic performance** (i.e., without using your actual name but only your ClassID) into the roster of academic performance of all students in this class? In this case, other students (your classmates) will be able to see your (without your actual name but anonymized ClassID) academic performance in this course	43%	43%	14%	**2.29 (86%)**
(5) Do you want the instructor to **identify the up-to-date levels of students' analytical, technical, management, and communication skills** in this course?	71%	29%	0%	**2.71 (100%)**
(6) Do you want the instructor to **clearly identify current at-risk students** (without actual names of students but only anonymized ClassIDs), i.e. students who currently have very low academic performance (usually, at the level of a "D" grade and below)?	57%	29%	14%	**2.43 (86%)**
Diagnostic Analytics of LA-based pedagogy ("Why did it happen?")				
(7) Do you want to know why you have a **certain identified overall current score** (i.e. total obtained points) in this course so far?	57%	29%	14%	**2.43 (86%)**
(8) Do you want to know why you have **certain identified current levels of analytical, technical, management and communications skills** in this course so far?	71%	14%	14%	**2.57 (86%)**
(9) Are you willing to **stop by the instructor's office during office hours** to discuss the above-mentioned two topics of Diagnostic Analytics?	43%	29%	29%	**2.14 (71%)**
(10) Did you try to stop by the instructor's office to **get help, get answers to your questions and/or improve your knowledge on specific topics** in this course?	57%	14%	29%	**2.29 (71%)**

(*continued*)

Table 3. (*continued*)

Questions to survey students	1	2	3	Av. GR score
(11) Did you try to **complete optional homework learning assignments** (for extra points) in this course?	86%	14%	0%	**2.86 (100%)**
(12) Did you try to **submit/provide your solution for optional in-classroom assignments**/discussions (for potential extra points) in this course?	71%	0%	29%	**2.43 (71%)**
Predictive Analytics of LA-based pedagogy ("What happened so far?")				
(13) Do you want **the instructor today to predict your total final score** (and, as a result, your final grade) in this course?	57%	14%	29%	**2.29 (71%)**
(14) Do you want to see or **get data about predicted final scores and/or final grades of other students in this course** under the condition that no student names but anonymized ClassIDs only will be associated with those data? In this case, you will be able to compare your predicted academic performance with the academic performance of your classmates	43%	43%	14%	**2.29 (86%)**
(15) Do you want the instructor to **include predicted data about your anonymized final score and/or final grade** (i.e. without using your actual name but only ClassID) **into the roster of predicted academic performance of all students** in this class? In this case, all other students (i.e. your classmates) will be able to see predicted academic performance of all students (but without actual name of any student but student ClassID only)	29%	57%	14%	**2.14 (86%)**
(16) Do you want the **instructor today to predict levels of your analytical, technical, management, and communication skills** by the end of this course?	43%	43%	14%	**2.29 (86%)**
(17) Do you want the instructor today to **predict and show to all classmates** (without actual names of students but their ClassIDs only) **potential students who are predicted to be at-risk** by the end of this course (i.e. students with predicted low-level academic performance by the end of this course)?	29%	57%	14%	**2.14 (86%)**
(18) Do you want **to receive anonymous and confidential alerts via email if the Smart Learning Analytics (SLA) system identifies you as currently an at-risk student** or predicts you to be an at-risk student by the end of this course?	43%	57%	0%	**2.43 (100%)**

(*continued*)

Table 3. (*continued*)

Questions to survey students	1	2	3	Av. GR score
Prescriptive Analytics of LA-based pedagogy ("How can we make it happen?")				
(19) Do you want to be able to **set up a new goal** (for example, a new desired final grade or final score in this course by the end of semester) based on current outcomes of Descriptive and Diagnostic Analytics, and **simulate with the Smart Learning Analytics (SLA) system potential needed activities/achievements/scores for remained learning assignments**? (In other words, Are you comfortable getting the answer to the following question: Can I achieve that new goal by the end of this course or not?)	100%	0%	0%	**3.00 (100%)**
(20) Do you want **the instructor and/or tutor to provide additional guidance and/or recommendations on what you should do** to achieve your new academic goal (for example, new final grade in this course) and/or improve your academic performance accordingly (based on the calculated prediction) in this course?	86%	14%	0%	**2.86 (100%)**
(21) Are you willing to **spend more than recommended 3 h per week of independent student work with a tutor** to improve your current academic performance in this course?	43%	29%	29%	**2.14 (71%)**
In general, what is your opinion about LA-based pedagogy	**61%**	**25%**	**14%**	**2.52 (86%)**

Based on obtained and presented data in the tables,

(1) A total of 79.83% of 23 undergraduate and 9 graduate students surveyed had a positive opinion about LA-based pedagogy, including "I like it very much" – 54.38% and "I somewhat like it" – 25.45%;

(2) the average score of all undergraduate and graduate students' answers regarding LA-based pedagogy is equal to a high score of 2.37 on 3-point scale (where 3.00 represents "I like this teaching strategy very much" student opinion).

It is interesting to compare the obtained average score with the average student scores of other analyzed modern teaching strategies – the outcomes are presented in our previous work [18]:

- *Learning-by-Doing* teaching strategy: 2.63 average score by both undergraduate and graduate students;
- *Flipped Classroom* teaching strategy: 2.30 average score by both undergraduate and graduate students;

- *Collaborative Learning* teaching strategy: 2.18 average score by both undergraduate and graduate students;
- *Gamification of Learning* teaching strategy: 2.12 average score by both undergraduate and graduate students.

5 Conclusions and Future Steps

Conclusions. The described on-going multi-aspect research, design and development project at the InterLabs Research Institute at Bradley University (IL, USA) is aimed at active use of systematic approach to identify, analyze, design, develop, test and, eventually, implement and recommend various components of SmP, i.e. a set of innovative teaching strategies that strongly support (a) SmE, SmU, and SmC concepts, and (b) smartness levels and smart features of SmE.

The obtained research findings and design and development outcomes enabled us to make the following conclusions:

1. Students really like LA-based pedagogy as a part of SmP. A total of 79.83% of 23 undergraduate and 9 graduate students surveyed had positive opinions about LA-based pedagogy, including 54.38% of "I like it very much" answers and 25.45% of "I somewhat like it".
2. The average score of 23 undergraduate and 9 graduate students surveyed regarding LA-based pedagogy is equal to high score of 2.37 on 3-point scale.
3. In accordance with student feedback, LA-based pedagogy enables students to take control of their own learning process and academic performance in a course; it also improves the quality of transparency of student academic performance to students.
4. Students were very impressed by the quality and depth of opportunities of different parts of LA – descriptive, diagnostic, predictive and prescriptive analytics.
5. Students appreciated information about status and prediction levels of their analytical, technical, communication and management skills in a course.
6. LA-based pedagogy improves (a) quality of teaching by faculty, (b) monitoring of student academic progress and performance in a course, (c) identification of at-risk student and on-time intervention by faculty to improve student performance, (d) prediction of student final score and grade, and (e) quality of identification of student skills in a course – analytical, technical, management, and communication.
7. Students expressed their concerns about (a) privacy and ethical aspects of LA in cases when anonymized data of a particular student (even with no actual student name but student ClassID) is displayed to all students in the class, and (b) quality of algorithms used in predictive analytics to forecast possible student final score and/or grade in a course.

Next Steps. Based on (a) obtained research data, findings and outcomes, and (b) developed and tested SmP components, the future steps in this research, design and development project are:

1. Increase quality of supporting Smart Learning Analytics system it terms of data automation (data cleaning, data processing and data visualization), and prediction of student final score and/or final grades in a course.
2. Perform summative and formative evaluations of LA-based pedagogy by local and remote students as well as instructors in additional Computer Science and Computer Information Systems courses such as Advanced Topics in Software Engineering, Web and Mobile Software Systems, Computer Architecture, Mobile Technology, Software and Computer Security, Integrative Programming, Game Design and Development, Game Engine Design and Modification, and Advanced Programming.
3. Identify SmP requirements for a smart teacher – a person who will be qualified to teach courses in modern SmC environment, including a set of requirements to technical skills and innovative teaching strategies (i.e. components of SmP) to be actively and smoothly used by a faculty in a modern SmC.

Acknowledgements. The authors would like to thank Mr. Ashok Shah, Mr. Tim Krock, Ms. Pravallika Vemulapalli, Mr. Cade McPartlin and Mr. Nicholas Hancher - the research associates of the InterLabs Research Institute and graduate and undergraduate students of the Department of Computer Science and Information Systems (CS&IS) at Bradley University - for their valuable contributions to this research, design and development project.

We also would like to thank Dr. Steven Dolins, Professor and Chair of the CS&IS Department for his long-term strong support of our research in SmE, SmU and SmP areas.

This research project is partially supported by grant REC # 1326809 at Bradley University (2015–2018).

References

1. Arroway, P., Glenda M., O'Keefe, M., Yanosky, R.: Learning Analytics in Higher Education. Research report. Louisville, CO: ECAR, March 2016. https://library.educause.edu/resources/2016/2/learning-analytics-in-higher-education
2. Sclater, N., Peasgood, A., Mullan, J.: Learning Analytics in Higher Education, A review of UK and international practice, April 2016. https://www.jisc.ac.uk/sites/default/files/learning-analytics-in-he-v2_0.pdf
3. Learning Analytics For Tracking Student Progress, Hanover research, November 2016 https://www.imperial.edu/research-planning/7932-learning-analytics-for-tracking-student-progress/file
4. Friesen, N.: Learning analytics: readiness and rewards. Can. J. Learn. Technol. (2013). http://learningspaces.org/wordpress/wp-content/uploads/2013/05/Learning-Analytics1.pdf
5. Society for Learning Analytics Research Available. https://solaresearch.org/
6. Siemens, G.: he journal of learning analytics: supporting and promoting learning analytics research. J. Learn. Anal. 1(1), 3–4 (2014)
7. Suchithra, R., Vaidhehi, V., Iyer, N.E.: Survey of learning analytics based on purpose and techniques for improving student performance. Int. J. Comput. Appl. 111(1), 22–26 (2015)
8. Borkar, S., Rajeswari, K.: Attributes selection for predicting students' academic performance using education data mining and artificial neural. Network 86(10), 25–29 (2014)

9. Tempelaar, D.T., Cuypers, H., van de Vrie, E., Heck, A., van der Kooij, H.: Formative assessment and learning analytics. In: Proceedings of the 2013 International Conference on Learning Analytics and Knowledge (LAK), April 8–12, 2013, Leuven, Belgium, (2013). https://pdfs.semanticscholar.org/29db/489b4532ee7e983a17f5653d65b31e9cb93d.pdf

10. Uskov, V.L., Bakken, J.P., Howlett, R.J., Jain, L.C. (eds.): Smart Universities: Concepts, Systems, and Technologies, 421 p. Springer, Cham (2018). ISBN 978-3-319-59453-8, https://doi.org/10.1007/978-3-319-59454-5

11. Uskov, V.L., et al.: Smart pedagogy for smart universities. In: Uskov, V.L., Howlett, R.J., Jain, L.C. (eds.) Smart Education and e-Learning, pp. 3-16. Springer, Cham (2017). ISBN 978-3-319-59450-7. https://doi.org/10.1007/978-3-319-59451-4

12. Uskov, V.L., Howlett, R.J., Jain, L.C. (eds.): Smart Education and e-Learning 2017, 498 p. Springer, Cham (2017), ISBN 978-3-319-59450-7. https://doi.org/10.1007/978-3-319-59451-4

13. Uskov, V.L., et al.: Smart university taxonomy: features, components, systems. In: Uskov, V.L., Howlett, R.J., Jain, L.C. (eds.) Smart Education and e-Learning 2016, pp. 3–14, Springer, Cham (2016). ISBN 9783319396897. https://doi.org/10.1007/978-3-319-39690-3

14. Uskov, V.L., Howlett, R.J., Jain, L.C. (eds.): Smart Education and Smart e-Learning, 512 p., Springer, Cham (2015). ISBN 9783319198743. https://doi.org/10.1007/978-3-319-19875-0

15. Neves-Silva, R., Tsihrintzis, G.A., Uskov, V.L., Howlett, R.J., Jain, L.C. (eds.): Smart Digital Futures, 808 p., IOS Press, Amsterdam, June 2014, ISBN 978-1614994046

16. Uskov, V., Bakken, J., Pandey, A.: The Ontology of Next Generation Smart Classrooms. In: Uskov et al. (eds.) Smart Education and Smart e-Learning, 510 p., pp. 3–14 Springer, Cham (2015). ISBN 978-3-319-19874-3

17. Uskov, V.L. Bakken, J.P. et al.: Building smart learning analytics system for smart university. In: Uskov, V.L., Howlett, R.J., Jain, L.C. (eds.) Smart Education and e-Learning 2017, pp. 191–204. Springer, Cham (2017). ISBN 978-3-319-59450-7. https://doi.org/10.1007/978-3-319-59451-4

18. Uskov, V.L., Bakken, J.P. et al.: Smart pedagogy: innovative teaching and learning strategies in engineering education. In: Proceedings of the 2nd IEEE EDUNINE International Conference on Engineering Education, IEEE, Buenos-Aires, Argentina, March 2018

19. Burlea Schiopoiu, A., Burdescu, D.D.: The development of the critical thinking as strategy for transforming a traditional university into a smart university. In: Uskov, V., Howlett, R., Jain, L. (eds.) Smart Education and e-Learning 2017, Smart Innovation, Systems and Technology, Vol. 75, pp. 67–74. Springer, Cham (2017)

20. Burlea Schiopoiu, A., Burdescu, D.D.: An integrative approach of E-Learning: from consumer to prosumer. In: Uskov, V., Howlett, R., Jain, L. (eds.) Smart Education and e-Learning 2016, Smart Innovation, Systems and Technology, vol. 59, pp. 269–279. Springer, Cham (2016)

Procedural Medical Training in VR in a Smart Virtual University Hospital

Håvard Snarby[1], Tarald Gåsbakk[1], Ekaterina Prasolova-Førland[2(✉)],
Aslak Steinsbekk[3], and Frank Lindseth[1,4]

[1] Department of Computer and Information Science, NTNU Norwegian
University of Science and Technology, Trondheim, Norway
`hsnarby@gmail.com, tarald.gaasbakk@gmail.com,`
`frankl@idi.ntnu.no`
[2] Department of Education and Lifelong Learning, NTNU Norwegian
University of Science and Technology, Trondheim, Norway
`ekaterip@ntnu.no`
[3] Department of Public Health and General Practice, NTNU Norwegian
University of Science and Technology, Trondheim, Norway
`aslak.steinsbekk@ntnu.no`
[4] SINTEF Medical Technology, Trondheim, Norway

Abstract. A smart virtual university hospital representing a real-life one, can prepare students for direct patient contact and provide possibilities for clinical practice. Such a virtual hospital will support student learning by providing adaptive and flexible solutions for practicing a variety of clinical situations at the students' own pace. This paper builds on a previous research on the Smart Virtual University concept and explores the possibilities for medical procedural training with Virtual Reality. A scenario focusing on pre-operative neurosurgical procedural training was developed and tested. The procedural training has been enhanced with real world medical data (MRI and ultrasound). The feedbacks from the tests have been generally positive, both in terms of general user experience and expected learning outcomes. The paper presents the implementation procedure, evaluation results and outlines the directions for future work.

Keywords: Smart virtual university hospital · Medical procedural training
Virtual Reality

1 Introduction

A smart virtual university hospital mirroring a real life one can prepare students for direct patient contact and provide opportunities for improving clinical practice. We have worked with the long-term idea of establishing an online Virtual University Hospital (VUH) to create a holistic system and a venue for learning, research, and development [1]. The idea is to make a virtual mirror of the St. Olav's University Hospital (St. Olav), which is integrated with the Faculty of Medicine and Health Science at the Norwegian University of Science and Technology (NTNU) in Trondheim, Norway. St. Olav is a modern university hospital with a state of the art technological platforms and modern clinical buildings. The first prototype of the Virtual

© Springer International Publishing AG, part of Springer Nature 2019
V. L. Uskov et al. (Eds.): KES SEEL-18 2018, SIST 99, pp. 132–141, 2019.
https://doi.org/10.1007/978-3-319-92363-5_12

University Hospital has been used for interprofessional communication and collaboration training [1].

Many studies have reported on the potential of 3D virtual worlds for educational activities, especially in medicine and health care [2, 3]. With affordable Virtual Reality (VR) technology, virtual environments can be used in combination with motion tracking and head-mounted displays (HMD, VR goggles) to increase the sense of immersion, making it more believable and transferable to the real life [4–6].

The VR experiences include surgical simulators [7–10], general training facilities for nurses and doctors [2, 3, 6] and anatomy education [4]. The literature reports several benefits in terms of the learning outcomes and student satisfaction [5, 6]. This teaching method can be cost-efficient and user-friendly alternative to real-life role-plays and training programs [2, 3, 10].

In medicine and health care there is a long history of physical simulations with many of the same benefits that VR simulation can offer, such as being safe, available outside the clinical setting, a controlled learning environment and providing opportunity for repeated training. However, traditional simulation tends to be expensive and located at centers with highly specialized equipment, often meaning restricted access [7, 11]. VR simulation has therefore a number of advantages in procedural training that are have been explored in the literature, e.g. for [8–10].

There have been attempts to develop affordable surgical simulators using general purpose input devices. E.g. at Japitur College of Engineering in India, Mathur [12] has developed a surgeon trainer using Oculus Rift with the low-cost input device Razer Hydra [13]. The users of the systems are using the headset to view the virtual world, and the controllers to interact with it from a stationary sitting position.

Accessing medical data such as MR and ultrasound is essential at all phases of medical treatment, from diagnosis to surgery. An example of a simulator supporting virtual interaction with medical data is Dextroscope [14], made to support surgical evaluation and decision making. The system uses preoperative images in combination with segmentation of critical anatomic structures to present an information-fused 3D model on a stereographic display.

Another example of a system that is used at multiple hospitals in the United States, is a NeuroTouch neurosurgical VR simulator [11], consisting of two parts, NeuroTouch Plan and NeuroTouch Sim. In the NeuroTouch Plan images from Computed Tomography (CT) and Magnetic Resonance Imaging (MRI) are used to create 3D models of the human head, allowing the neurosurgeons to examine the brain for preparation of neurosurgery by defining the surgical corridor. Once it is defined, the user is moved to the NeuroTouch Sim part, where he/she performs the surgery. However, this system is very complex and expensive, and it is therefore need for more research in this area.

While real world medical data could have immense benefits in a VR learning environment, it has been found to be hard to implement [15]. Also, more knowledge about adding medical data into VR simulators is needed. In order to address this gap in the related work, the main goal of this project has been to develop a low-cost solution for room-scale Virtual Reality (Oculus Rift and HTC Vive) with integrated medical data for pre-operative procedural training. Detailed description of the scenarios can be found in the master thesis that this article is based on [16]. The work is part of the Smart University Hospital framework earlier described by the authors [1].

An application focusing on pre-operative neurosurgery was developed, in collaboration with the specialists. The rest of the paper is organized as follows: the following section outlines the implementation details and training scenarios. Section 3 presents the results of a preliminary evaluation with medical specialists using the neuro-scenario, while Sect. 4 discusses the findings. Section 5 concludes the paper, outlining directions for future work.

2 Methods

2.1 Implementation of Real World Medical Data

The models for the environment like hospital hallway, rooms and operating rooms were based on the original design of the VirSam project [1]. They were modelled by taking photographs of the operating room at the Neurological center at St. Olavs, in order to create a familiar environment for neurosurgeons. Unity3D was chosen as the platform for the development of the solutions, to be used with both HTC Vive or Oculus Rift.

Moving on from VirSam [1], it was important to improve the interaction with the virtual world. In the original gynecological VirSam scenario, when performing ultrasound, the physician could only look at a frozen screen and vocally declare the findings. In this new development, two different approaches were taken to implement real world medical data into the application.

The first approach was to map ultrasound images to the position of the ultrasound probe in the virtual world, meaning the image on the virtual ultrasound machine updates when the user moves the probe. The second approach used an advanced software (Volume Viewer Pro) for converting MRI data to a 3D model. This approach was used to create an interactable panel consisting of the axial, coronal and sagittal (ACS) view of MRI data, enabling users to interact with the data like how they do in real life. There are two sliders on each panel, which the user can drag to change the images. It is also possible for the user to interact with the panel by pointing a finger on the screen.

Another implementation of the second approach is the use of the navigation wand. The navigation wand gives the user the ability to navigate the MRI data based on the wand's position relative to the patient's head. The data is displayed on an ACS panel over the surgical bed.

2.2 Interaction in the Virtual World

There are several items the user needs to interact with to complete the tasks in the scenario. It was decided to keep the number of interactable buttons to a minimum to enable new users to adapt quickly to the sensation of VR. The first button (tracker button) located under the thumb of the user, is used to teleport/transport the user to the desired destination. The second button (trigger) is used for interacting with objects in the virtual world. This button is located under the index finger of the user. Whenever the user hovers an object with their virtual hands, the hovering controller will be visible as translucent yellow, indicating which button the user must press to pick up the object.

The button on the controller will produce a small vibration. The vibration of the controllers is also used to indicate to the user that they are interacting with the virtual world. For instance, when the user points on the ACS panels or use the ultrasound system, the controller will vibrate.

3 Evaluation of the Neurology Scenario

3.1 Tasks Performed in the Scenario

The neurosurgery scenario consisted of four main tasks:

1. Select case. The users can play through three different cases, with different tumor positions. The first task they are given is to select a case they want to play.

2. Locate tumor based on MRI data. After selecting the case, the user gets the task of locating the tumor and mark it using the ACS panel. When the user is satisfied with the result, they can press the solution button to toggle between the correct solution and their own result (Fig. 1).

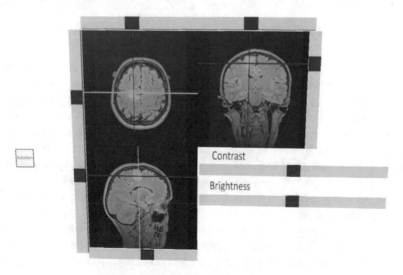

Fig. 1. Locating tumor on the ACS panel

3. Patient positioning, shaving and marking the entry point of the craniotomy. Based on the result obtained in the previous task, the user is tasked with positioning the patient correctly on the surgical bed. The head of the patient needs to be correctly rotated to ensure proper access to the tumor and optimal conditions for ultrasound acquisition. When the patient is laid correctly, the patient needs to be shaved to get access to the point of operation. When the patient is shaved, the user marks the point of surgical entry. After marking the point of entry, the user clicks on the solution button to toggle between the correct setup and their own.

3. Navigation and ultrasound acquisition. When the patient is positioned correctly, the next task is to verify the position of the tumor. This is done in two steps Both steps are illustrated in Fig. 2. The first step is verification by use of the navigation system (MRI data on the ACS panel above the bed).

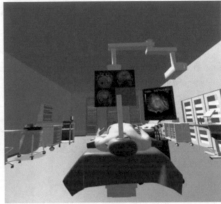

Fig. 2. Use of navigation system and ultrasound

The second step is verification by ultrasound. In neurosurgery, it is not sufficient to rely solely on the MRI when performing an operation. Firstly, the registration of the patient's MRI data is not completely accurate. The inaccuracy of the registration process can be close to a centimeter. In addition, once the operation has started, the brain might move around, making the tumor move as well. To remedy these issues, ultrasound is used complementary to the MRI during the operation as this modality is not affected by registration issues. Being an intraoperative modality, the ultrasound images will give an updated view of the current patient anatomy. New ultrasound scans are performed when needed to determine the current position of the tumor and critical brain tissue. These scans are compared with each other throughout the surgery. In the last task, the user has to perform the first ultrasound scan. To perform the scan, the user picks up the ultrasound probe and move it to the point of entry on the head. If the tumor in observed at the assumed location the user has completed the scenario (Fig. 2).

3.2 Testing Scenario

Test of First Prototype. The first round of testing was performed during the 2017 Network Conference for medical simulation in Norway, held at the Medical Simulation (MedSim) Centre at St. Olavs hospital spring 2017. The version tested was a prototype, meaning that it was not yet robust enough to leave the testers on their own. In order to make new users familiar with the unusual interface, 'virtual hands' were implemented and programmed to animate in a manner that simulated the physical movement of the fingers.

This test was conducted using one VR headset Oculus Rift, i.e. one person at the time did the testing. The individuals participating in these tests were 12 technical and medical experts from different parts of Norway, interested in simulation in medical practice. They were only asked to try out the functionality and fill out a short survey. All questions were posed in a manner that should be answered on a Likert scale from 1, strongly disagree, to 5 strongly agree (Fig. 3).

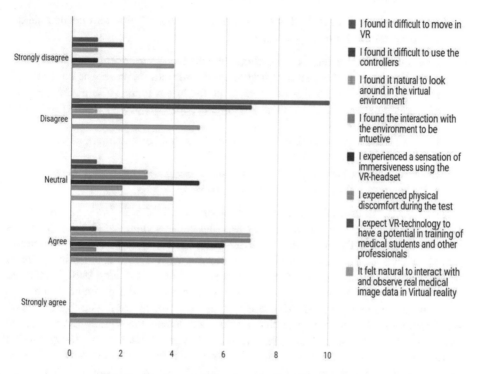

Fig. 3. Chart created based on feedback from MedSim

Test of Final Version. In June 2017, the application presented in this paper was demonstrated at the 9th international Ultrasound in Neurosurgery course in Trondheim, Norway. The application was used as one of four stations were the participants should explore the simulation methods and get a first-hand experience with different equipment and concepts regarding ultrasound in neurosurgery. The focus in this test was shifted somewhat from what was already established in earlier user tests. Instead of focusing on whether the users were able to interact with the VR gear and the virtual environment, observations were made on how the users interacted with the medical equipment and the medical data.

Setup and Execution. The execution was a three-step process for each test subject. Firstly, to give the testers an overview of the application and make them ready for the tasks they were given a quick walk-through by team members. Secondly, they were given a quick introduction to the controller and the different buttons they needed to use to interact with virtual world. Lastly, they were given the headset and guided through the scenario.

Test Subjects. The test subjects in this execution were mainly surgeons in training or surgeons who wanted to expand their knowledge on ultrasound guided neurosurgery. The age of the participant ranged from late twenties to sixties. None of the testers had any previous experience in the use of VR.

Findings. The data gathered in this test was made by observation and posttest talks with focus on the following aspects:

(1) How quickly the testers became familiar with the environment (number of seconds from start of the test until first teleport and until the case was chosen)
(2) How quickly the testers were able to locate the tumor in the ACS-panel; (number of seconds between the first attempt to interact with the panel and until the tumor was marked correctly)
(3) How well the testers were able to position the patient and mark him correctly (related to the pre-defined correct solution)
(4) How well the testers were able to interact with the medical data using the MR-Wand and ultrasound probe and to locate the tumor.

The average time users spent before choosing the case was 1:24 min, while the average time required to locate the tumor was just under a minute. The time spent was mostly used as an indicator on the user-friendliness of the system, without comparing to how much time is required for such procedures in real life. Placing the patients correctly proved mostly to be successful. Only two out of the nine participants had crucial errors in their placements or markings. The rest of the participants had the patient mostly correctly placed. All those who got the time to test the navigation wand and ultrasound were able to use those tools and locate the tumor. Some needed assistance on how to use one, or both, but most understood that they were supposed to use the VR gear as they would do in real life. The main issue with the ultrasound probe was the inability to rotate the probe and get a proper rotated view, but most testers did not notice this flaw. One of the testers claimed that the navigation system and the ultrasound implementations made it feel like he was working on a real patient. The user did not look at the patient when performing the navigation, instead he was only looking at the screen.

4   Discussion

The application was tested on two occasions, both during and after the development was finished. The feedback gathered presents interesting findings related to the usefulness of the application, especially in an educational context.

Interactivity. Developers of VR solutions need to account for typical problems experienced by complete novices, such as understanding how to aim the teleport beam and what button to click are problems. The fact that users need some time to become familiar with the VR gear is to be expected. As shown in the test performed at MedSim, there was largely a consensus that the interaction with the controllers was not difficult. During earlier tests, some users had trouble finding out which items were possible to interact with and which button to press to interact with it. Therefore, the interactable items have been differentiated from non-interactable ones by shape and color.

VR-sickness. A recurring concern with VR applications have been VR-sickness, which may leave users unable to use applications for a prolonged period of time. In this project, numerous design decisions have been adopted to minimize the sensation of nausea and strain on the users, such as optimizing performance and implementation of teleport functionality, which removes the discrepancy between the user's physical and virtual movement. As a result, only one participant reported discomfort during the test (Fig. 3).

Using Real Medical Data. From the literature, the use of real world medical data is perceived as having a significant potential in educational applications, but difficult to implement [15]. The evaluation of the prototype shows that it is indeed possible to utilize real world medical data to support a relatively low-cost learning application. Proper use and interpretation of MR and ultrasound images is a field of study in itself. Therefore, it is paramount that the interaction with the medical data is as similar to the real world interaction as possible, or manipulated in a way as intuitive as possible. From the user tests, we can see that the average tester spent an average of just under a minute to properly locate the tumor in the ACS-Panels. Having the ability to use the sliders and finger pointing, the ACS-Panels were made to simulate desktop computers and touch screens in the operating room, enabling the testers to interact with the equipment in a way that was familiar to them.

At the same time, being able to interact with the models and imagery of the human body in a fully immersive virtual environment enables exploration of the data in a new way. Utilizing medical imagery to recreate apparatus and scenes that are known and familiar to the users will help them to perform training on well-known procedures, as well as developing new training scenarios focusing on advanced interaction, interpretation and 3D visualization of medical data.

5   Conclusions and Future Work

In this paper, we have discussed developing a virtual university hospital environment designed for procedural training in VR. The initial setup and introduction to VR might be time consuming. However, with a well-designed user interface, and with careful guidance, it is achievable to have novice users interacting with the virtual world with some degree of proficiency after minutes. The design of educational applications is a complex process, and one way to solve this is to design such an application based on a set of tasks where each one brings educational value, preferably in collaboration with an expert. This is why the application in this project have been centered around scenarios with a clear structure and learning goals.

Integrating real world medical data further increased the sense of immersion the users could feel when playing the scenario and gave them an exciting new way to interact with MR and ultrasound images. The users were forced to use their knowledge of MRI and ultrasound in order to complete the given tasks properly through experimentation and analysis of the medical data in the virtual environment. Construction of these types of scenarios needs to be done in close collaboration with experts in the given field.

Based on experimentation and exploration in this project, multiple directions for future work have been identified:

Medical Data. The testers suggested that further integration of medical data in the application is desirable, with two possible improvements. The first one is the ability to slice the data at a random angle and offset, simulating the use of an ultrasound probe in the real world. Another way of interacting with the data that was suggested in the tests with surgeons, is the ability to view the navigation markers position on the ACS-Panel that is positioned above the bed. Such functionality will immerse the user even further as well as open for more complex scenarios.

Improved Feedback System. Today, the application provides an expert-set solution for the scenarios, and the user may assess his or her success in the scenario based on the solution. During the development, the team came up with different ideas for more a complex feedback systems, some sort of numerical feedback that in a meaningful way grades the performance of the user. A score system could be visible after the user is done setting up the patient, and could be showed in VR or outside, with pass or fail values, giving the users further feedback on what is expected from them.

An example of possible feedback to the user could be to inform about potential dangers to the patient if the entry point and target were marked in such a way that critical structures would be on the surgical path down to the tumor. Personalized and adaptive feedback based on different metrics like the placement of the head or the angle of the neck of the patient would be useful for the learner as well.

Further Development of the Smart Virtual University Hospital. The goal of this project was to develop a medical learning application for procedural training within the Smart Virtual University Hospital framework. Compared to the previous implementations of Virsam/VUH [1], the user was able to move freely in the virtual environment, interact with equipment using physical movements and access medical data. This contributed to increased immersion, made it more user-friendly and resulted in reduced time needed to be proficient in the application. The new features introduced such as smart medical equipment with medical images dynamically changing according to the user's hand movement constitute additional smartness levels, e.g. adaptation [17].

References

1. Prasolova-Førland, E., Steinsbekk, A., Fominykh, M., Lindseth, F.: Practicing interprofessional team communication and collaboration in a smart virtual university hospital. In: Uskov, V., Bakken, J., Howlett, R., Jain, L. (eds.) Smart Universities, Smart Innovation, Systems and Technologies, vol. 70, pp. 191–224. Springer (2018)
2. Wiecha, J., Heyden, R., Sternthal, E., Merialdi, M.: Learning in a virtual world: experience with using second life for medical education. J. Med. Internet Res. **12**(1) (2010). https://www.jmir.org/2010/1/e1/

3. Lowes, S., Hamilton, G., Hochstetler, V., Paek, S.: Teaching communication skills to medical students in a virtual world. J. Interact. Technol. Pedagogy **3**, e1 (2013)
4. Jang, S., Black, J.B., Jyung, R.W.: Embodied cognition and virtual reality in learning to visualize anatomy. In: Ohlsson, S., Catrambone, R. (eds.) 32nd Annual Conference of the Cognitive Science Society, Portland, OR, 12–14 August, pp. 2326–2331. Cognitive Science Society, Austin (2010)
5. Huang, H.M., Liaw, S.S., Lai, C.M.: Exploring learner acceptance of the use of virtual reality in medical education: a case study of desktop and projection-based display systems. Interact. Learn. Environ. **24**(1), 3–19 (2016)
6. Khanal, P., Gupta, A., Smith, M.: Virtual worlds in healthcare. In: Gupta, A., Patel, L.V., Greenes, A.R. (eds.) Advances in Healthcare Informatics and Analytics, pp. 233–248. Springer International Publishing, Cham (2016)
7. Stefanidis, D., Sevdalis, N., Paige, J., Zevin, B., Aggarwal, R., Grantcharov, T., Jones, D.B., Association for Surgical Education Simulation Committee & Association for Surgical Education Simulation Committee: Simulation in surgery: what's needed next? Ann. Surg. **261**(5), 846–853 (2015)
8. Lewis, T.M., Aggarwal, R., Rajaretnam, N., Grantcharov, T.P., Darzi, A.: Training in surgical oncology–the role of VR simulation. Surg. Oncol. **20**(3), 134–139 (2011)
9. Kühnapfel, U., Cakmak, H.K., Maaß, H.: Endoscopic surgery training using virtual reality and deformable tissue simulation. Comput. Graph. **24**(5), 671–682 (2000)
10. Pelargos, P.E., Nagasawa, D.T., Lagman, C., Tenn, S., Demos, J.V., Lee, S.J., Bui, T.T., Barnette, N.E., Bhatt, N.S., Ung, N., Bari, A.: Utilizing virtual and augmented reality for educational and clinical enhancements in neurosurgery. J. Clin. Neurosci. **35**, 1–4 (2017)
11. CAE Healthcare: Neurotouch (2017). https://caehealthcare.com/surgical-simulation/neurovr
12. Mathur, A.S.: Low cost virtual reality for medical training. In: 2015 IEEE Virtual Reality (VR), pp. 345–346. IEEE (2015)
13. Razer: Razer hydra. https://www2.razerzone.com/au-en/gaming-controllers/razer-hydra-portal-2-bundle
14. Shi, J., Xia, J., Wei, Y., Wang, S., Wu, J., Chen, F., Huang, G., Chen, J.: Three-dimensional virtual reality simulation of periarticular tumors using Dextroscope reconstruction and simulated surgery: a preliminary 10-case study. Med. Sci. Monit. Int. Med. J. Exp. Clin. Res. **20**, 1043 (2014)
15. Egger, J., Gall, M., Wallner, J., Boechat, P., Hann, A., Li, X., Chen, X., Schmalstieg, D.: HTC Vive MeVisLab integration via OpenVR for medical applications. PLoS One **12**(3), e0173972 (2017)
16. Snarby, H., Gåsbakk, T., Lindseth, F., Prasolova-Førland, E., Steinsbekk, A.: Medical Procedural Training in Virtual Reality (2017)
17. Uskov, V.L., Bakken, J.P., Pandey, A., Singh, U., Yalamanchili, M., Penumatsa A.: Smart university taxonomy: features, components, systems. In: Uskov, L.V., Howlett, J.R., Jain, C.L. (eds.) Smart Education and e-Learning 2016. pp. 3–14. Springer International Publishing (2016)

Learning English Grammar in the Smart Learning Environment

Ivana Simonova[✉][iD]

University of Jan Evangelista Purkyne, Usti nad Labem, Czech Republic
ivana.simonova@ujep.cz

Abstract. Smart education in the smart learning environment belongs to the fields which are rather new; therefore, researches are highly required. In this article results of the research which focused of the process of learning within the smart learning environment are presented. The learning content acquired by the students includes selected English grammar phenomena. The main objective of the research is to find out how students learn in the smart learning environment. Research sample consisted of 61 part-time bachelor students of Faculty of Informatics and Management, University of Hradec Kralove, Czech Republic. Data were collected in three phases: (1) face-to face monitoring the entrance knowledge before the process of learning started by pre-test, (2) autonomous learning supported by teacher's feedback in the smart environment and continuous testing by post-test1, (3) final face-to-face testing of acquired knowledge by post-test2. Results proved that the process of learning conducted in the smart learning environment brought some improvements, however, students had problems in application of advanced grammar phenomena in the active way.

Keywords: Smart learning environment · Higher education · English grammar
E-Learning · ESP · English for Specific Purposes

1 Introduction

Smart learning environment (SLE) belongs to the frequently discussed phenomena within the educational science. When considering the role of latest hardware and software within the process of instruction, a wide scale of approaches opens both to the teachers and learners how SLE can enhance the instruction. However, the question still exists why (on the basis of what features) a developer calls the educational product to be "smart" [1: 2732].

Within the process of development, the content of the word "smart" changed substantially. From the simple use of devices connected to the Internet, currently, the term "smart" has been used in the context of learning environment, educational technologies and conceptual framework [1]. We agree with Spector [1: 2731] stating the SLE is based on human intelligence, joining following features: knowledge, i.e. the access to information and ability to process it; task support, i.e. providing learners with information and tools to conduct and finish the task; learner and context sensitivity, i.e. offering the learner appropriate assistance in building and keeping motivation within

© Springer International Publishing AG, part of Springer Nature 2019
V. L. Uskov et al. (Eds.): KES SEEL-18 2018, SIST 99, pp. 142–150, 2019.
https://doi.org/10.1007/978-3-319-92363-5_13

the task solving; and reflection and feedback, i.e. considering learner's progress and the whole process of acquiring new knowledge.

The term "smart education" describes learning in the digital age, when learners are expected to meet the needs of the work and life in the 21st century. Moreover, the term "context-aware ubiquitous learning" is often used to describe the process exploiting mobile devices, wireless communication and sensing technologies to enable learners to interact with both real-world and digital-world objects, i.e. within smart learning students learn from the real world by using digital resources [2]. Within this context, the interactions between learners and environments are included. Therefore, smart learning environments can be deemed the technology-supported learning environments that implement adaptations and provide appropriate support in the right places and at the right time on the basis of individual learners' needs. These needs may be determined by examining learning behaviors, performance, and the online and real-world contexts in which learners are situated [3].

The adjective "smart" in smart learning involves some characteristics similar to those attributed to a person who is regarded as being "smart". Some of these characteristics include the ability to "adapt in creative and innovative ways to novel or unusual circumstances, to engage in appropriate planning prior to making a decision or taking an action and in doing things that are generally effective and efficient" [4: 2]. In other words, being 'smart' is attributed to "an action or decision that involved careful planning, cleverness, innovation, and/or a desirable outcome" [4: 2]. The effectiveness of the SLE results in desirable learning outcomes which are better compared to those reached under other conditions; the efficiency relates to the cost effectiveness, i.e. to the fact that not significantly higher expenses are required to support and maintain the smart learning environment compared to a non-smart conditions.

Another definition of "smart" is provided by the Interactive Technology and Smart Education Information peer-reviewed journal. It states that "SMART" is used as an acronym referring to interactive technology that offers a more flexible and tailored approach so as to meet diverse individual requirements by being "Sensitive, Manageable, Adaptable, Responsive and Timely" to educators' pedagogical strategies and learners' educational and social needs' [5].

As discovered by Uskov et al. [6: 4], 76% of teachers state the technology allows them to respond to a variety of learning styles, 77% of teachers declare the technology used in the class motivates students to learn, 74% of administrators say the digital content in schools increases student engagement, 91% of administrators emphasize the effective use of educational technology is critical to their mission of high student achievement. Instead of others, Uskov et al. analyzed strategies which are also included in our concept of work within SLE, e.g. working with e-books; giving students personalized assessment to support the learning process; exploiting the BYOD (Bring Your Own Device) approach; solving problems through computational thinking gained within studying other subjects from the field of computing; crossover learning connecting formal and informal learning; personal inquiry learning, i.e. applying collaborative inquiry and active investigation in learning; analytics of emotions, i.e. responding to the emotional state of students; learning to learn; and coincidental learning etc. [6: 5–6].

Moreover, as several authors state, e.g. [4, 7, 8], the SLE is required not only to be effective, efficient and engaging, but also adaptive from the view of individual's learning preferences [9]. It should support students' motivation [10] and provide smart guidance through the learning process [11]. Thus we can conclude that when referring to having knowledge and ability to accomplish difficult matters, to being innovative and to responding appropriately to specific circumstances and challenges, the criteria are identical, either we have technologies, or people in mind [1].

Reflecting the above mentioned, the main objective of this article is to introduce and research the concept of learning English grammar within the smart learning environment.

2   Process of Learning English Grammar in ESP: State of Art at the Faculty of Informatics and Management

Four courses (one per semester) of English for Specific Purposes (ESP) are taught in the part-time bachelor study programmes for the students of Applied Informatics and Information Management. Upper secondary school graduates are required to have the knowledge of English on the B1 level of CEFR (Common European Framework of Reference for Languages) [12: 217]; in spite of the fact, this pre-condition is not met by many of them. Nevertheless, they are enabled to start university studies. However, they are expected to fulfil the missing knowledge as soon as possible. Therefore, they attend private courses and lessons, and their effort is also supported within the ESP1 course, the learning content of which focuses on English grammar so as the weak students could reach the B1 level as minimum. Then, other courses targeted at further knowledge and skills follow: in ESP2 the skill of reading comprehension of professional texts is developed, ESP III focuses on written communication, and in ESP IV the oral communication and presentations are trained. Before graduation, students are required to reach the B2 level of CEFR.

For this purpose, both the face-to-face and remote types of learning are exploited, which means the blended approach is applied. Within each semester, which is 12 weeks long, 24 face-to-face hours are conducted, been held in four six-hour blocks in the classroom. After each block remote autonomous learning within the smart learning environment Blackboard is required. As defined above, not all learning environments can be called "smart". However, Blackboard, which was originally designed as the learning environment, and if exploited appropriately, meets all the requirements necessary for conducting the process of teaching/learning, and works as the smart learning environment.

3   Methodology

3.1   Research Sample

Students of the Faculty of Informatics and Management, University of Hradec Kralove, Czech Republic, enrolled in Applied Informatics and Information Management part-time bachelor programmes and studying the subject of ESP1 participated in the research (N = 61). They were from 19 to 44 years old: 48 respondents (78.7%) were in the interval of 24–36 years; ten respondents (16.4%) were of female gender. However, neither of these variables was considered within the research. Students' level of general English knowledge was defined according to the CEFR [12: 217], considering both the results of the entrance exam test and respondents' self-assessment: the research sample consisted of 14 students of A1 level, 28 students of A2, 17 students of B1 and 2 students of B2 level. Additionally, six students working in English-speaking companies proved their knowledge of English by C1 certificates (four students) and C2 certificates (two students). They had an individualized learning content in ESP I subject, which reflected their higher-level of knowledge; therefore, they were not included in the research sample.

3.2   Research Objective

Reflecting the fact, the smart education is widely exploited and accepted, the question appears among others, whether learning results developed within the smart environment are better. In other words, whether students learn more, or learn more easily, or in a shorter time period. In this research, attention was paid to the field of acquiring the learning content.

The main objective of this research is to discover how much students learn, when acquiring selected phenomena of English grammar, if the process of learning is enhanced by the smart learning environment.

3.3   Research Method, Tools and Hypotheses

The process of learning English grammar was conducted in three phases.

In the first phase, an electronic list of 44 grammar phenomena (both in English and Czech language) was given to the students on their first face-to-face ESP lesson before the learning content started to be explained and acquired. The list included grammar phenomena e.g. irregular nouns, comparative and superlative forms of adjective and adverbs, modal verbs, tenses, infinitive structures, gerunds, question tags, "have something done" structure, so/neither am I, etc. Students' task was to write a short and simple sentence containing the phenomenon in the appropriate context. Samples from general English, not ESP were required. The time period for completing the list was 45 min. After the lesson, students submitted the list to the SLE Blackboard and teacher considered the sample sentences counting one point per each correct sample (maximum score was 44 points). This result is called the pre-test score further on. Moreover, from the very beginning students were acknowledged their work and effort is purposeful, finally targeting to improving their knowledge, but the way how they will show the teacher what they learned was unknown to them at that time.

In the second phase, students were to read books relating to their field of study and/or work, i.e. to Applied Informatics and Information Management, and to find the required grammar phenomena in professional texts. Once detected, the sentence containing the appropriate grammar phenomenon including the reference was inserted in the list. Both printed and e-sources could be used. Before, during and after creating the list of referred sentences the students were allowed and encouraged to exploit various learning aids, e.g. the presentation created by the teacher which provided the summary of all required grammar phenomena with a few samples, any grammar books or student's books with exercises and the key, web pages relating to learning English, printed and e-dictionaries etc. They could conduct discussions with other students, both in the SLE or social networks, and consider the in/correctness of single sentences, or share sources and methods of searching for single phenomena. Most of the students exploited electronic sources from the Internet and their work-related texts. As the students of Informatics and related fields, they were also able to use advanced search engines. Totally, 2,684 sentences should have been submitted. Unfortunately, from the reason of teacher's inadvertence eight sentences were missing. Out of the real amount of 2,676 sentences 82% of them (2,194 sentences) were collected from e-sources, i.e. from the Internet, Intranet of companies students worked for, CD-ROMs etc., where search engines could have been exploited and students could have searched for the grammar phenomena by using part of the structure (e.g. "have been" for Present Perfect Simple, or Present Perfect Continuous tenses). By the way, this approach was the main source of incorrect samples; however, this is not the topic of this research. The time for completing the list was six weeks. Then, each student submitted it through the SLE Blackboard to the teacher for evaluation. Identically to the pre-test, one point per each correct sample was scored (maximum score was 44 points); this result is called the post-test1 score further on. The teacher provided feedback to each student where correct and incorrect sentences were marked and further explanations provided, including the links to the summarizing presentation mentioned above. The deadline for sending the feedback was one month before the end of semester as minimum. Advanced students completed the list of sentences faster and the first come first served principle in providing the feedback was applied by the teacher. Then, student's task was to study the analysis and if needed, to contact the teacher for further support so as to acquire the learning content as deep as possible.

In the third phase, which was conducted within the final face-to-face credit test at the end of the semester, students were given the same list of grammar phenomena as in previous phases. Again, their task was to write simple sentences relating to each grammar phenomenon; in this test the use of professional vocabulary in the sentences was required. Identically to the pre-test and post-test1, one point per each correct sample was scored (maximum score was 44 points). This result is called the post-test2 score further on.

So as to reach the main research objective, i.e. to discover how much students learn when acquiring selected phenomena of English grammar if the process of learning is enhanced by the smart learning environment, two hypotheses were set:

H1: There exists the statistically significant difference between pre-test and post-test1 scores.

H2: There exists the statistically significant difference between post-test1 and post-test2 scores.

4 Results

Data collected in the pre-test, post-test1 and post-test2 were processed by appropriate statistic methods. Results are structured into two parts: (1) descriptive statistics and (2) verification of hypotheses.

4.1 Descriptive Statistics

Results of descriptive statistics are displayed in Table 1, presenting the values of total amount of respondents (N), Mean, Standard Deviation (SD), Minimum (Min) and Maximum (Max) score, score range (Range), Median, Mode and results of two tests of normality data distribution (Shapiro-Wilk W test and Kolmogorov-Smirnov test). As clearly seen mainly in Mean values, the pre-test score was 29.3 points. After six-week long study period in the SLE, the post-test1 score increased to 35.6 points. However, in the final post-test2 the score decreased to 30.1 points (maximum score was 44 points in each test). The normality of data distribution was rejected by both the statistic tests in all three phases (pre-test, post-test1, post-test2). Reflecting this result, non-parametric tests were applied for verification of hypotheses.

Table 1. Descriptive statistics.

Heading level	Pre-test	Post-test1	Post-test2
N	61	61	61
Mean	29.34426	35.60656	30.14754
SD	8.54963	6.535234	8.835225
Min	4	18	7
Max	42	44	43
Range	38	26	36
Median	32	37	31
Mode	-	37	30
Normality: Shapiro-Wilk W	0.9203044 (R)	0.9287397 (R)	0.9445432 (R)
Normality: Kolmogorov-Smirnov	0.1699123 (R)	0.1581928 (R)	0.1162891 (R)

R: Reject normality

4.2 Verification of Hypotheses

First, the paired difference for pre-test score and post-test1 score was calculated by Wilcoxon Signed Rank test. Reaching the Z-value = 6.6112 ($\alpha = 0.05$; probability level = 0.000000), the first hypothesis H1 was verified. This result means that statistically significant difference was discovered between the pre-test and post-test1 scores (see Fig. 1).

Second, the paired difference for post-test1 score and post-test2 score was calculated by Wilcoxon Signed Rank test. Reaching the Z-value = 3.6336 ($\alpha = 0.05$; probability level = 0.000280), the second hypothesis H2 was verified. This result

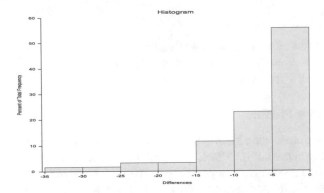

Fig. 1. Differences in test scores: pre-test versus post-test1.

means that statistically significant difference was discovered between the post-test1 and post-test2 scores (see Fig. 2).

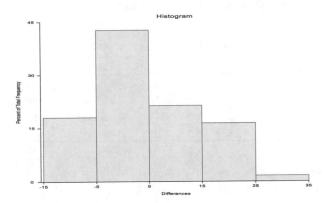

Fig. 2. Differences in test scores: post-test1 versus post-test2.

5   Discussions and Conclusions

Despite differences in paired tests with both hypotheses were rather high (Z-value for H1 was 6.6112; Z-value for H2 was 3.6336), the Mean test score decreased from 35.60656 points in post-test1 to 30.14754 in post-test2. This result shows that students' level of knowledge in the field of grammar phenomena, when they were to prove they had acquired both the terminology and relating samples, slightly increased after studying in SLE from 29.34426 points in pre-test. However, students were not able to produce appropriate sentences actively, without the possibility to exploit various learning aids provided in the SLE (e.g. search engines, printed and electronic grammar books and dictionaries etc., as listed above).

Discussion over these results is rather difficult, mainly because of the fact neither the identical, nor similar approach has been applied yet. However, the smart learning, education and concepts have been exploited by numerous universities, to some extent. As stated by Uskov et al. [13], who had overviewed innovative approaches in the smart university area, some projects and researches in this field are being conducted. Learners' progress within the smart learning environment was evaluated by Elhoseny et al. [14]. They considered the structure of smart teaching methods and proved the general conclusion that innovations should work in a viable, proficient, adaptable and continuous way. They proposed the Higher Particle Optimization towards triggering learning engagement activities. On the other side, as analyzed by Klimova [15], teacher's role is still important, even though it was changed within the work in the smart learning environment. The study revealed that despite living in the e-society, some teachers were not able to effectively implement latest technologies in their teaching. In other words, the continuous training is (and will be) needed in the future. A new paradigm called e-learning ecosystem was designed by Ouf et al. [16]. It is described as a teacher-student model applying a fixed pathway suitable for all learners through exploiting smart features.

When considering the proposed solutions, it is difficult to decide whether any of them can work for efficient teaching and learning of foreign languages, particularly in the field of learning, or practising English grammar. The solution presented in the research produced statistically significant results in the second phase, which was conducted in the smart learning environment. Within future research activities, students' performance could be investigated and compared in the experimental and control groups with the smart learning environment as the variable. Finally, it should be stated we are aware of the limits of our research, particularly the size of research sample, its structure which do not allow to generalize the results. Moreover, as the "smartness" of learning environment will be increasing, new features should be monitored which will bring a more complex view.

References

1. Spector, J.M.: Smart learning environments: concept and issues. https://www.researchgate.net/publication/301612985_Smart_Learning_Environments_Concepts_and_Issues. Accessed 09 Jan 2018
2. Mikulecky, P.: Smart environments for smart learning. In: Capay, M., Mesarosova, M., Palmarova, V. (eds.) 9th International Scientific Conference on Distance Learning in Applied Informatics, DIVAI 2012, pp. 213–222. UKF, Nitra (2012)
3. Ossiannilsson, E.: Challenges and opportunities for active and hybrid learning related to UNESCO Post 2015. In: Handbook of Research on Active Learning and the Flipped Classroom Model in the Digital Age (2016). https://www.igi-global.com/chapter/challenges-and-opportunities-for-active-and-hybrid-learning-related-to-unesco-post-2015/141011. Accessed 30 Dec 2017
4. Spector, J.M.: Conceptualizing the emerging field of smart learning environments. Smart Learn. Environ. 1(1), 1–10 (2014)
5. Interactive Technology and Smart Education, Aim and scope. http://www.emeraldgrouppublishing.com/products/journals/journals.htm?id=itse. Accessed 10 Jan 2018
6. Uskov, V.L., Bakken, J.P., Penumatsa, A., Heinemann, C., Rachakonda, R.: Smart pedagogy for smart universities. Smart Innov. Syst. Technol. 75, 3–16 (2017)

7. Spector, J.M.: Foundations of Educational Technology: Integrative Approaches and Interdisciplinary Perspectives, 2nd edn. Routledge, New York (2015)
8. Sims, R.: Rethinking (e)learning: a manifesto for connected generations. Distance Educ. **29** (2), 153–164 (2008). https://doi.org/10.1080/01587910802154954. Bates, T., Spector, J.M., Merrill, M.D. (Eds.): Special issue: effective, efficient and engaging (E3) learning in the digital age
9. Kumar, V., Kinshuk, S.G.: Causal competencies and learning styles: a framework for adaptive instruction. J. e-Learning Knowl. Soc. **7**(3), 13–31 (2011)
10. Kim, C.: The role of affective and motivational factors in designing personalized learning environments. Educ. Tech. Res. Dev. **60**(4), 563–584 (2012)
11. van Merriënboer, J.J.G., Kirschner, P.A.: Ten Steps to Complex Learning: A Systematic Approach to Four-Component Instructional Design, 2nd edn. Routledge, New York (2013)
12. Common European Framework of Reference for Languages: learning, teaching and assessment. https://www.coe.int/en/web/common-european-framework-reference-languages/table-1-cefr-3.3-common-reference-levels-global-scale. Accessed 30 Dec 2017
13. Uskov, V.L., Bakken, J.P., Howlett, R.J., Jain, L.C.: Innovations in smart universities. Smart Innov. Syst. Technol. **70**, 1–7 (2017)
14. Elhoseny, H., Elhoseny, M., Abdelrazek, S., Riad, A.M.: Evaluating learners'progress in smart learning environment. In: Advances in Intelligent Systems and Computing, vol. 639, pp. 734–744 (2018)
15. Klimova, B.: Smart teacher. Innovations in smart universities. Smart Innov. Syst. Technol. **75**, 321–329 (2017)
16. Ouf, S., Abd Ellatif, M., Salama, S.E., Helmy, Y.: A proposed paradigm for smart learning environment based on semantic web. Comput. Hum. Behav. **72**, 796–818 (2017)

In Re Launching a New Vision in Education and e-Learning: Fostering a Culture of Academic Integrity in e-Learning

Michele T. Cole[✉], Daniel J. Shelley, and Louis B. Swartz

Robert Morris University, Moon Township, PA 15108, USA
{cole, shelleyd, swartz}@rmu.edu

Abstract. This paper presents the results of two studies conducted in 2017 on students' use of technology, specifically social media, to enhance their e-learning experience. The first study focused on how students had used social media for e-learning. The second shifted the focus from the students' own experiences to their observations on all students' use of social media. Both studies emanated from earlier research on students' perceptions of academic integrity and the use of technology in the online learning environment. One hundred eight-five graduate and undergraduate students participated in the first survey. One hundred sixty graduate and undergraduate students participated in the second. Based on enrollment status and academic level, there were statistically significant differences in both studies with regard to the use of Facebook and Snapchat to enhance students' learning experience. There were differences between the two studies when the questions focused on the student's individual use or all students' use, of social media to cheat. In the first, students admitted to having used one or more social media to cheat "a few times" (3.85%–6.15%). In the second, more students said that other students used social media to cheat "a few times" (10.53%–21.05%). Based on enrollment status and academic level, there were statistically significant differences in the second study on using technology to cheat. Responses to the final questions concerning the student's own feelings about academic dishonesty and what actions he/she might take to stop others from cheating did not evidence a culture of academic integrity.

Keywords: Academic integrity · Ethics · e-Learning · Social media Technology

1 Introduction

The International Center for Academic Integrity [1] defines academic integrity as a commitment to six fundamental values on which ethical behavior rests: honesty, trust, fairness, respect, responsibility, and courage. Universities such as the one represented in this study promote a culture of academic integrity in a variety of ways. These include posting the academic integrity (AI) policy prominently on its website, enforcing student codes of conduct, emphasizing the importance of academic integrity in activities with incoming students (including international students), and asking faculty to reinforce the

© Springer International Publishing AG, part of Springer Nature 2019
V. L. Uskov et al. (Eds.): KES SEEL-18 2018, SIST 99, pp. 151–164, 2019.
https://doi.org/10.1007/978-3-319-92363-5_14

concepts in syllabi, as well as to enforce the policy in their courses. Despite these efforts, research has found that students continue to commit acts of academic dishonesty [2–8].

Ethics and academic integrity intersect with technology, specifically with regard to the Internet and the challenges posed by the students' use and misuse of it [9]. It is impossible to talk about the students' use or misuse of the Internet without considering the continually evolving technology that supports smarter e-learning, including social media [10].

The ease of committing acts of academic dishonesty is enhanced by the technology widely available on the Internet. Sendag et al. [2] highlight e-dishonesty – the use of online resources to facilitate unethical behavior - in their investigation of extent of student involvement in academic dishonesty practices at their university.

In comparing the use of social media in academic dishonesty by Canadian and Chinese students, Hernandez [3] notes the inevitability of students' using social media in e-learning. Ward and Wolf-Wendel [11] remarked that the use of social media in higher education was ubiquitous. "We live in a society where social media use is universal…social media are now important spaces on the higher education landscape… how students use social media …to enhance their college experiences is of critical interest…" [11, p. 7].

Selwyn [12] notes that the growth of social media has transformed how the Internet is used. It is no longer merely a source of information, with applications like Facebook, Twitter, and YouTube. The Internet also allows students to share and collaborate, making it difficult for higher education to set parameters for its use for purposes of academic integrity.

There have been a number of studies of academic dishonesty in both the online and on-ground learning environments [13–15]. Some have focused on the widespread occurrence of plagiarism [16]. Others have noted the difficulties administrators and academics in higher education have experienced when trying to foster a culture of academic integrity in their schools [17, 18]. McNabb and Olmstead [19] surveyed faculty. Cole et al. [20], Miller et al. [21], and Thakkar and Weisfeld-Spolter [22] surveyed students.

In a Swedish study on academic dishonesty, ethics and learning, Colnerud and Rosander [4] observed that the students in their study evaluated academic dishonesty on a scale tied to the level of difficulty of the course, not on the act of academic dishonesty itself. Interestingly, with the exception of explicitly unfair behavior, the authors argue that if learning results from academic dishonesty, it can be morally justified. If learning is the goal, it has been achieved - the student has done his duty. If the student has learned the material, the grade is fairly earned. If the student has learned the material, the expected competency might have been achieved.

In an earlier study [6], researchers found that academic dishonesty was viewed differently by students in the e-learning and traditional classroom environments as well. Using technology, social media, to learn course material was not viewed as cheating, but rather as in sync with the "real world."

Comparing the attitudes and actions of Asian and European students in a New Zealand study, Henning et al. [7] found little difference between the two groups with regard to academically dishonest behaviors and in reporting cheating. In contrast,

Hernandez [3] found that in her sample of Canadian and Chinese students that the Chinese group scored higher with regard to instances of academic dishonesty using social media than the Canadian group. She concluded that institutions need to be aware of the differences in students' perceptions of academic dishonesty. Presumably, awareness of the differing perceptions would carry over to institutional polices on academic integrity and how such institutional policies might be communicated.

In a study of nursing students' attitudes toward reporting dishonest behavior, Tayaben [5] found that students would be unlikely to take responsibility for enforcing a culture of academic integrity. Sendag et al. [2] found that the incidence of e-dishonesty was significantly greater among first-year college students than it was for graduate students. They argue for early intervention to help to create a culture of academic integrity.

With the proliferation of smart technology, learners are presented with significant opportunities to enhance their experience in what Uskov et al. [23] refer to as the next-generation smart classroom. The question here is: How can institutions and instructors use smart e-learning to counter attitudes toward academic dishonesty? How can smart e-learning help to foster a culture of academic integrity? This study is in part an investigation of that question.

2 The Study

The study examined students' use of social media to enhance e-learning and/or to gain an unfair advantage, to cheat, or otherwise act dishonestly when completing course-work. The study is a continuation of the authors' investigation of the use of technology and its effect on academic integrity in online instruction [6, 13]. The purpose of the research is to determine how a culture of academic integrity can be fostered in e-learning. To establish a baseline, participants were asked how they felt about academic dishonesty and what would they do to stop it.

2.1 Research Questions

1. Do students use social media to enhance learning?
2. Do students use social media to cheat?
3. Do students believe that academic dishonesty is wrong?

2.2 Methodology

Researchers developed a 17 question survey in *Question Pro*, a web-based survey instrument that is supported by the University. A solicitation to participate in the surveys was posted in the course shells in Blackboard for selected graduate and undergraduate courses in the School of Education and the School of Business in the spring 2017 (7 April 2017 – 16 May 2017) and fall 2017 (22 September 2017 – 25 November 2017) terms. Participation was voluntary and responses were anonymous. Results from each set of surveys were transferred from *Question Pro* to SPSS for analysis. Results are reported in the aggregate.

2.3   Sample

The sample from the first study in spring 2017 included undergraduates in on-ground and online education and business courses (EDUC2200, BLAW1050) and graduate students in on-ground and online education and business courses (EDUC6130, EDUC6140, HRMG 6250 and HRMG 6900).

The sample from the second study in fall 2017 also included undergraduates in on-ground and online business and education courses (BLAW 1050, EDUC 2000, EDUC2200, ECED 4000) and graduate students in on-ground and online business courses (HRMG 6350 and MBAD6010). Each of the on-ground courses included an online component.

In the first study, undergraduates represented more than 80% of the respondents. Of the 157 undergraduates who participated, 76 (38.97%) were freshman; 55 participants were sophomores representing 28.21% of the undergraduate sample. There were 13 (6.67%) juniors and 13 (6.67%) seniors who participated in the survey. Thirty-eight (19.49%) of the respondents were graduate students.

The number of undergraduates and graduate students participating was more even in the second study. More than 46% of the students were undergraduates. Of those, 25 (25.62%) were sophomores; 21 (13.12%) were juniors; and 17 (10.62%) were seniors. The remaining 11 students (6.88%) were freshman. Of the 83 graduate students, 82 (51.25%) were master's level students; one respondent was a doctoral candidate.

One hundred eighty-five of the 195 students who began the spring 2017 survey completed it for a completion rate of 94.87%. All of the 160 students who began the fall 2017 survey completed it for a completion rate of 100%. Eighty-eight percent of the participants in the spring survey were full-time students - 12% were part-time students. In the fall 2017 survey, 65% of the participants were full-time students - 34% were part-time students. Forty-seven percent of the respondents in the spring study were male - 53% were female. In the second study, 52.5% of the participants were male - 47.5% were female. Undergraduates made up 80.51% of the sample in the first study. Graduate students represented 19.49%. In the second study, undergraduates represented 46.24% of the sample and graduate students represented 51.87% of the sample. A question on age range was included in the second survey. The majority, 62.50% were between the ages of 18–24, followed by 26.88% who were between the ages of 25–34. Thirteen respondents were between the ages of 35–44. Table 1 presents the participant profile for each study. Not all participants responded to every question. Some students in the first study responded to the demographic questions, but did not complete the survey.

Table 1. Participant profiles

Demographics	Spring 2017 study (N = 185)	%	Fall 2017 study (N = 160)	%
FT/PT	172/23	88.21/11.79	104/55	65.0/34.38
M/F	91/104	46.67/53.33	84/76	52.50/47.50
UG/G	157/38	80.51/19.49	74/83	46.24/51.87
Freshman	76	38.97	11	6.88
Sophomores	55	28.21	25	15.62
Juniors	13	6.67	21	13.12
Seniors	13	6.67	17	10.62
Graduate Level	38	19.49	83	51.87[a]
18–24			100	62.50
25–34			43	26.88
35–44			13	8.12

[a]Three students self-reported as "Other"

2.4 Procedure

A 17 question survey was developed and administered in spring 2017. An additional question on age range was added to the fall 2017 survey. The first three questions in the first study asked for enrollment status, gender, and academic level. The fourth question in the second study asked for respondents' age range. The next three questions focused on the use of social media - Facebook, Twitter, and Snapchat/Instagram - to enhance learning. There were six questions asking if students had used social media, Facebook, Twitter, Snapchat/Instagram, or technology such as smart phones and features such as texting and screen shots, to gain an advantage, to cheat or to act dishonestly. The final two questions asked participants to evaluate cheating or other acts of academic dishonesty and asked how they would respond to others who were cheating. There were three open-ended questions at the end asking for additional comments.

In the first study, students were asked if they had used social media to enhance learning and/or to gain an unfair advantage or to cheat. The questions were reframed in the second study to ask if respondents knew if students in general used social media to enhance and/or to gain an unfair advantage or to cheat. Feedback from the first study indicated that students might be more forthcoming if the questions were not directly related to their own behavior.

Selected data from the completed surveys were transferred into SPSS from *Question Pro* for analysis. Independent samples t-tests were conducted on the three questions that focused on the use of specific social media to enhance learning and on the following six questions asking if specific social media and certain technology was used to gain an advantage, to cheat or to act dishonestly based on enrollment status, gender, and academic level.

2.5   Results

Spring 2017 Study

RQ 1. Do Students Use Social Media to Enhance Learning? In response to the questions asking if respondents used social media to enhance learning or to support academic endeavors, 21.54% said they did use Facebook "somewhat," 13.85% used Twitter "somewhat," and 13.08% used Snapchat/Instagram "somewhat." Almost 47% said that they used Facebook, but not for college-related activities. More than 47% of those responding said that they used Twitter, but not for college related activities. More than 60% said that they used Snapchat/Instagram, but not for college-related activities. Those who said they never use Facebook represented 28.21% of the sample. Those who never used Twitter represented 34.62% of the sample. Almost 20% (19.23%) said that they never used Snapchat or Instagram to enhance learning.

RQ 2. Do Students Use Social Media to Cheat? When asked if the student had ever used Facebook, Twitter, and/or Snapchat and Instagram to gain advantage or to cheat, overwhelmingly, the responses were "never." One hundred twenty-two students (93.85%) said that they never used either Facebook or Twitter to cheat. One hundred twenty students (92.31%) said that they never used Snapchat or Instagram to cheat.

When asked if the student had ever used a smartphone to share answers on an exam, 72.31% said "never," and 24.62% said, "a few times." Responses were similar to the questions on texting answers or using screen shots to share answers to an exam. More than 70% said "never" to texting to sharing information on an exam, and 73.85% said "never" to using screen shots to share information on an exam. Almost 25% said they had used texting to share answers on an exam and 21.54% said they had used screen shots to share answers to an exam.

Fall 2017 Study

RQ 1. Do Students Use Social Media to Enhance Learning? In response to the questions asking if respondents knew if students used social media to enhance learning or support academic endeavors, 26.25% said they knew that students used Facebook, 16.67% said students used Twitter, and 31.58% said that students used Snapchat/Instagram somewhat to enhance learning. Thirty percent said that they knew that students used Facebook, but not for college-related activities. More than 47% of those responding said that they knew that students used Twitter, but not for college related activities. Almost 22% said that they knew that students used Snapchat/Instagram, but not for college-related activities. Those who said they did not know of anyone using Facebook to enhance learning represented 38.12% of the sample. Those who said they did not of anyone using Twitter represented 32.46% of the sample. Approximately 14% said that they did not know of anyone who used Snapchat or Instagram to enhance learning.

RQ 2. Do Students Use Social Media to Cheat? The next questions asked if the respondent knew if students used Facebook, Twitter, and/or Snapchat/Instagram to gain an advantage or to cheat. Most responded "never" to Facebook (82.46%), to Twitter (86.84%), and to Snapchat/Instagram (72.81%). With regard to students sometimes using social media to gain an unfair advantage or to cheat, respondents said they did know that students sometimes used Facebook (15.79%), and/or Twitter (10.53%), and/or Snapchat/Instagram (21.05%) to gain an unfair advantage or to cheat.

When asked if the respondent knew if students used smartphones to share answers on an exam, 43.86% said "a few times," and 44.74% said "never". Responses were similar to the question on texting answers to an exam. More than 43% said they knew that students sometimes texted to share answers to an exam, and 43.86% responded "never" in response to the question. When asked if they knew that students used screen shots to share exam answers, 31.58% of the respondents said "a few times," and 55.26% said "never." Not all students responded to each question. Table 2 compares the findings.

Table 2. Comparison of results on social media and learning

Using social media to enhance learning	Sp.17 (N = 185)	%	Fall 17 (N = 160)	%
Facebook: Somewhat/Not for College-Related Activities/Never	42/91/55	21.54/46.67/28.21	42/48/61	26.25/30.00/38.12
Twitter: Somewhat/Not for College-Related Activities/Never	18/62/45	13.85/47.69/34.62	19/54/37	16.67/47.37/32.46
Snapchat/Instagram: Somewhat/Not for College-Related Activities/Never	17/83/25	13.08/63.85/19.23	36/25/16	31.58/21.93/14.04

Independent samples t-tests were conducted on each of the nine questions with regard to using social media and related technology to enhance learning and/or to gain an advantage or to cheat based on enrollment status, gender and academic level. Statistically significant differences were found between full-time and part-time students in the spring 2017 study with regard to the student's own use of Twitter and Snapchat/Instagram to enhance learning at the .05 level for Twitter (.030, equal variances assumed) and for Snapchat/Instagram at the .01 level (.008, equal variances assumed.) There were statistically significant differences based on gender with regard to the use of Twitter to enhance learning at the .05 level (.034, equal variances assumed.) Statistically significant differences were also found between sophomores and juniors with regard to the use of Facebook to enhance learning at the .05 level (.044, equal variances assumed.) There were no statistically significant differences with regard to students using social media or technology to gain an unfair advantage or to cheat found in the spring 2017 study (Table 3).

Table 3. Comparison of results on social media and cheating

Using social media to cheat	Sp. 17 (N = 185)	%	Fall 17 (N = 160)	%
Facebook to Cheat- Never/A Few Times	122/6	93.85/4.62	94/18	82.46/15.79
Twitter to Cheat - Never/A Few Times	122/5	93.85/3.85	99/12	86.84/10.53
Snapchat/Instagram to Cheat - Never/A Few Times	120/8	92.31/6.15	83/24	72.81/21.05
Smartphone to Cheat - Never/A Few Times	94/32	72.31/24.62	51/50	44.74/43.86
Texting to Cheat - Never/A Few Times	92/32	70.77/24.62	50/50	43.86/43.86
Screen Shots to Cheat - Never/A Few Times	96/28	73.85/21.54	63/36	55.26/31.58

In the fall 2107 study, statistically significant differences were found between full-time and part-time students with regard to the use of Facebook to enhance learning at the .05 level (.05, equal variances assumed.) There were statistically significant differences found between freshman and sophomores on the use of Snapchat/Instagram to enhance learning at the .05 level (.013, equal variances assumed.) Statistically significant differences were found between full-time and part-time students with regard to the use of texting to cheat at the .01 level (.000, equal variances not assumed.) There were also statistically significant differences between freshman and sophomores on the use of smartphones and texting to cheat at the .05 level (.017, equal variances assumed and .031, equal variances assumed respectively.) There were no statistically significant differences based on gender. Tables 4 and 5 display the results of the independent samples t-tests.

Table 4. Use of social media to enhance learning

Social Media	Study	Grouping variable	n	M	t	Sig. (2-tailed)
Facebook	Spring	Junior/Soph.	13/55	1.77/2.35	−2.052	.044
	Fall	FT/PT	104/55	2.15/1.82	1.972	.050
Twitter	Spring	FT/PT	113/17	1.96/1.47	2.188	.030
		M/F	59/71	2.07/1.75	2.138	.034
Snapchat/Instagram	Spring	FT/PT	113/17	2.10/1.59	.949	.008
	Fall	Fresh./Soph.	7/20	3.71/2.50	2.689	.013

RQ 3. Do Students Believe that Academic Dishonesty Is Wrong? Students were also asked what their personal feelings were toward academic dishonesty and what they would do if they observed others cheating. Responses to the spring 2017 and fall 2017 surveys were similar. Sixty students or 46.15% in the first survey answered that

Table 5. Use of Technology to Cheat

Social media	Study	Grouping variable	n	M	t	Sig. (2-tailed)
Smart phones	Fall	Fresh./Soph.	7/20	1.29/2.20	−2.553	.017
Texting	Fall	FT/PT	68/31	1.88/1.32	3.934	.000
		Fresh./Soph.	7/20	1.29/2.25	−2.280	.031

cheating or academic dishonesty is always wrong and unethical. Sixty-one or 53.51% in the fall survey said that academic dishonesty is always wrong and unethical. Forty students or 30.77% in the first survey responded that they did not really care, others cheat, but they did not. Thirty-seven (32.46%) in the second survey said that they did not really care, others cheat, but they did not. In the spring survey, 16 (12.31%) had no opinion. In the fall survey, 11 students (9.65%) had no opinion.

Many students would do nothing if they observed cheating. Sixty-one (46.92%) in the spring survey, and 38 (33.33%) in the fall survey said that they would do nothing if they observed another student in the course cheating. Fifteen (11.54%) in the spring survey responded that they would let the person cheating know and that they disapproved. In the fall survey, that number was 25 (21.93%). Twenty-nine (22.31%) had no opinion in the spring, and 36 (31.58%) had no opinion in the fall survey. Twenty-four (18.46%) responded that they would inform the instructor in the spring compared with 14 (12.28%) in the fall survey. Table 6 presents the results to the third research question. Not all students responded to each question.

Table 6. Beliefs about Academic Dishonesty and Consequences

Q: Academic Dishonesty	Spring 2017 study (N = 185)	%	Fall 2017 study (N = 160)	%
Always Wrong	60	46.15	61	53.51
Don't Care	40	30.77	37	32.46
No Opinion	16	12.31	11	9.65
Ok If not Caught	0	0	2	1.75
Ok Sometimes	14	10.77	3	2.63
Q: Action				
Ignore	61	46.92	38	33.33
Disapprove	15	11.54	25	21.93
No Opinion	29	22.31	36	31.58
Threaten	1	.77	1	.88
Inform	24	18.46	14	12.28

The final questions were open-ended asking for additional comments on the survey questions, academic integrity, and the area of research. There were 15 comments submitted in the spring survey and 23 comments submitted on the fall survey. These were illustrative of the grey areas between accessing widely available information for legitimate purposes and misusing the tools available to violate principles of academic integrity. For example:

- "I don't think a lot of students use social media to cheat, but I know 99% of students google answers to quizzes and exams that are online. There's nothing stopping them." [Spring 2017 Comment # 6].
- "I think this is a more difficult subject for online students where they can get help from others or the internet/their phones without it being noticed by anyone. I am much more willing to seek outside help during an online exam than I am in a classroom." [Spring 2017 Comment # 11].
- "I think there's a gray area between enhancing learning and violating academic integrity. For instance, if someone does not understand how to complete an assignment, they may call or text a classmate who does know how to do it. This classmate would share their work, while explaining to the confused student the processes and what to do, actively instructing the other student. While one student did make use of another's work, I don't see this as cheating or a violation of academic integrity since the student whose work was being used collaborated and helped the other student understand it for themselves." [Fall 2017 Comment # 3].
- "I have seen some Facebook friends post questions regarding a case, research paper, and/or discussion from their class to get more insight on issue or opinions from others that most likely may very from their own. I cannot say how this person used the information they were given and if it violated any of the academic rules." [Fall 2017 Comment # 9].
- "I believe social media is a great tool to enhance learning of concepts related to things in class but not necessarily 'pure theoretical textbook material'… while most students adhere to the principles of academic integrity there are some who routinely abuse this through the use of their smart phones. It becomes difficult to want to report these things when you also maintain friendships with these students outside of class. But, as a student who does adhere to strong personal ethics when completing work, it's very frustrating to see someone achieve the same results without any of the hard work. It will take time for technology to catch up to this problem." [Fall 2017 Comment # 13].

Limitations. The study was limited to students in two schools in one university in southwestern Pennsylvania. Surveys were conducted as part of undergraduate and graduate courses. While student participation was voluntary, extra credit was provided for participation.

3   Discussion

This study had a dual purpose: (1) to determine whether students felt that social media enhanced e-learning and whether students used, or knew that social media was used to cheat, and (2) to determine how students felt about acts of academic dishonesty. Results indicated that students did use social media such as Facebook, Twitter, Snapchat and Instagram to help them learn course material, but only to a limited degree. Results support findings by Neier and Zayer [24] from their study of students' perceptions and experiences with social media as a learning tool in the classroom, but differ in degree from Piotrowski's review of studies of students' views of the use of social media platforms [25].

Neier and Zayer found students to be cautious about which tools they used and how those tools were used for learning. "Overall, respondents indicated that they would be motivated to use social media in the classroom because it aligns with their desire to be interactive" [24, p. 9]. Students in this study also noted their use of social media primarily to connect with others. "I use social media to connect with other individuals in my industry to learn new skills, attend webinars and network…" [Fall 2017 Comment #10]. "I do not think that social media gives students an unfair advantage in my opinion as it is mostly used for social interaction not academics" [Fall 2017 Comment # 16]. Students' voluntary comments in this study raised questions as to how social media such as Facebook, Twitter, and Instagram could be used to support learning objectives.

Results from this study differed from the Baetz et al. [26] study of students' perceptions of academic integrity and cheating behaviors. In that study, only 7.5% of 412 students reported that they had not cheated. Less than 4% (3.9%) said they had cheated just one time. In this study, researchers found reported instances of, and/or observations of the use of social media and technology to gain an unfair advantage or to cheat to be limited. There were a few responses to the open-ended questions that suggested otherwise. "On many occasions I saw a classmate using their cell phone during a test. So the one time I reported it to the professor. Butbthat [sic] was only once, and I've seen this quite a bit but not with just that one student" [Spring 2017 Comment # 10]. "I think that more college students share answers to exams through the use of screen shots, text messages, and Google" [Spring 2017 Comment # 21].

Responses to questions on how students felt about academic dishonesty and what they might do to prevent could not be said to evidence a strong culture of academic integrity. While almost 50% (49.18%) of the respondents from the combined studies said that any type of cheating or academic dishonesty was always wrong and unethical, fewer than 16% (15.57%) would inform the instructor. More than 40% (40.57%) would do nothing - ignore it. Sixty-five students (26.63%) from the combined studies had no opinion about others cheating.

How can institutions and researchers use smart e-learning to counter the attitudes toward academic dishonesty? Whitley and Keith-Spiegel proposed four elements of what they called an academic integrity ethos: institutional integrity, a learning-oriented environment, a values-based curriculum, and an honor code [27]. Institutional integrity is a shared value, one that, through its visible efforts at promoting academic integrity, (including the institution of a student honor code) that the University demonstrates. As student-centered learning is one of the cornerstones of the University's strategic plan, again, one can argue that the University promotes a learning-centered environment.

In an earlier study [13], the researchers asked students in three graduate classes to discuss what academic integrity meant to them and how technology affected academic integrity. Responses demonstrated a belief that academic integrity was important, particularly in e-learning as resources were unlimited in the online environment. Related to the question of how instructors and institutions can foster a culture of academic integrity, these students remarked on the relationship between student and instructor. They also noted that while technology does make it easier for the student to cheat, it also makes it easier for the instructor to detect cheating. Here, the instructor plays a key role in using technology to help prevent intentional and unintentional instances of academic dishonesty. In the end though, students in that study felt that while technology certainly played a role, it was the student who was ultimately responsible for maintaining academic integrity.

As in the earlier study, researchers here suggest that to foster a culture of academic integrity, institutions and instructors need to reassess how academic integrity and academic dishonesty are perceived by all participants in e-learning if only to acknowledge the realities of a cyber-world. It appears that there are different views of how learning occurs. If students are accessing whatever resources they feel they need to learn the material, is it possible for smart e-learning to turn this "new reality" into a positive learning experience? Suggestions from earlier studies [12] include developing measurements for how well students are using technology to enhance their learning; restructuring course assignments to take advantage of resources available on the Internet, and enabling students to make appropriate use of those resources. Others have suggested adding different assessment strategies such as online collaborations as well as projects to assess learning while educating students on ethical behaviors [28].

IT has changed both how people cheat and how cheating may be detected. Additionally, students' view of what constitutes cheating has changed over time, with the result that what instructors consider to be cheating is not always identified as such by their students [29].

Smart e-learning continues to develop, driven by advances in technology and enhancements to pedagogy. Searching for the right mix of technology and pedagogy, O'Connell [30] refers to digital information ecology – a remix of technology, devices, data repositories and retrieval, networks, communication and information sharing, including social media. It is information ecology, she states, that is central to creating a culture of academic integrity. While information ecology has opened the door to opportunities for dynamic exchanges, it also has "extended the disciplinary and pedagogical challenges relevant to learning design and the broader institutional context of responsibility for academic integrity" [30, p 3]. She suggests a collaborative approach to curriculum development that explicitly teaches academic integrity as the enabler of participatory learning. Others [31] have proposed using drama as a strategy for fostering academic integrity.

Gamalel-Din [32] tells us that the vision of smart e-learning is to empower the student to learn and the instructor to teach. The challenge remains how to position technology to support a culture of academic integrity. Boehm et al. [33] argue that a key to fostering academic integrity is developing a philosophy that promotes academic integrity while it educates students and faculty on its principles, as opposed to emphasizing penalties for academic dishonesty. Others [34] suggest it may not be that simple. Faculty may agree that academic integrity is important without agreeing as to if or how it should be taught.

Despite recent protests from faculty on the value of online education [35, 36], it would appear that online instruction and e-learning are here to stay. Technology will continue to develop and students will take advantage of what the new technology has to offer. That being said, the manner in which a shared culture of academic integrity is communicated by institutions and instructors and, some may argue, fellow students is paramount.

References

1. ICAI: The fundamental values of academic integrity (2013). http://www.academicintegrity.org/icai/resources-2.php
2. Sendag, S., Duran, M., Fraser, M.R.: Surveying the extent of involvement in online academic dishonesty (e-dishonesty) related practices among university students and the rationale students provide: one university's experience. Comput. Hum. Behav. **28**, 849–860 (2012)
3. Hernandez, M.D.: Academic dishonesty using social media: A comparative study of college students from Canada and China. SAM Adv. Manag. J., **80**(4), 45–51 (2015)
4. Colnerud, G., Rosander, M.: Academic dishonesty, ethical norms and learning. Assess. Eval. High. Educ. **34**(5), 505–517 (2009)
5. Tayaben, J.L.: Attitudes of student nurses enrolled in e-learning course towards academic dishonesty: a descriptive-exploratory study. In: Saranto, K., et al. (eds.) Nursing Informatics, pp. 32–38 (2014). https://doi.org/10.3233/978-1-61499-415-2-32
6. Cole, M.T., Swartz, L.B., Shelley, D.J.: Students' use of technology in learning course material: is it cheating? Int. J. Inf. Commun. Technol. Educ. **10**(1), 35–48 (2014). https://doi.org/10.4018/ijicte.2014010104
7. Henning, M.A., Malpas, P., Manalo, E., Ram, S., Vijayakumar, V., Hawken, S.J.: Ethical learning experiences and engagement in academic dishonesty: a study of Asian and European pharmacy and medical students in New Zealand. Asia-pacific Educ. Res. **24**(1), 201–209 (2015). https://doi.org/10.1007/s40299-014-0172-7
8. Chapman, K.J., Davis, R., Toy, D., Wright, L.: Academic integrity in the business school environment: I'll get by with a little help from my friends. J. Mark. Educ. **26**(3), 236–249 (2004). https://doi.org/10.1177/0273475304268779
9. Hinman, L.M.: Academic integrity and the world wide web. Comput. Soc., 33–42 (2002). https://doi.org/10.1145/511134.511139
10. Klimova, B.: Assessment in smart learning environment- a case study approach. Springer International Publishing (2017). https://doi.org/10.1007/978-3-319-19875-0
11. Ward, K., Wolf-Wendel, L. (eds.): Social media in higher education: a monograph. In: Rowan-Kenyon, H.T., Martinez, A.M., Aleman and Associates. Social Media in Higher Education. ASHE Higher Education Report 42(5), Jossey-Bass: San Francisco (2016)
12. Selwyn, N.: Social media in higher education. The Europa World of Learning 2012 (2012). ISBN 978-1-85743-620-4
13. Cole, M.T., Swartz, L.B., Shelley, D.J.: Technology-infused education and academic integrity: are they compatible? In: Tomei, L. (ed.) Exploring the New Era of Technology-Infused Education. IGI Global, Hershey, Pa, pp. 241–262 (2017)
14. Ghaffari, M.: Instant gratification and culture of academic disintegrity: Implications of trinity paradigm of intelligence in developing a culture of integrity. Int. J. Interdiscip. Soc. Sci. **3**(10), 89–101 (2009)
15. McCabe, D.L., Trevino, L.K., Butterfield, K.D.: Cheating in academic institutions: a decade of research. Ethics Behav. **11**(3), 219–232 (2001). https://doi.org/10.1207/S15327019EB1103_2

16. Thomas, E.E., Sassi, K.: An ethical dilemma: talking about plagiarism and academic integrity in the digital age. Engl. J. **100**(6), 47–53 (2011)

17. Kidwell, L.A., Wozniak, K., Laurel, J.P.: Student reports and faculty perceptions of academic dishonesty. Teach. Bus. Ethics **7**, 205–214 (2003). https://doi.org/10.1023/A:1025008818338

18. McCabe, D., Pavela, G.: Some good news about academic integrity. Change, 32–38 (2000). https://doi.org/10.1080/00091380009605738

19. McNabb, L., Olmstead, A.: Communities of integrity in online courses: faculty member beliefs and strategies. J. Online Learn. Teach. **5**(2), 208–221 (2009)

20. Cole, M.T., Shelley, D.J., Swartz, L.B.: Academic integrity and student satisfaction in an online environment. In: Yang, H.H., Wang, S. (eds.) Cases on Online Communities and Beyond, pp. 1–19. IGI Global, Hershey (2013)

21. Miller, A., Shoptaugh, C., Wooldridge, J.: J. Exp. Educ. **79**, 169–184 (2011). https://doi.org/10.1080/00220970903567830

22. Thakkar, M., Weisfeld-Spolter, S.: A qualitative analysis of college students' perceptions of academic integrity on campus. Acad. Educ. Leaders. J. **16**, 81–88 (2012)

23. Uskov, V., Bakken, J.P., Pandey, A.: The ontology of next generation smart classrooms. Springer International Publishing (2017). https://doi.org/10.1007/978-3-319-19875-0

24. Neier, S., Zayer, L.T.: Students' perceptions and experiences of social media in higher education. J. Mark. Educ. **37**(3), 133–143 (2015)

25. Piotrowski, C.: Academic applications of social media: a review of peer- review research in higher education. Psychol. Educ. **52**, 15–22 (2015)

26. Baetz, M., Zivcakova, L., Wood, E., Nosko, A., De Pasquale, D., Archer, K.: Encouraging active classroom discussion of academic integrity and misconduct in higher education business contexts. J. Acad. Ethics **9**, 217–234 (2011)

27. Whitley, B.E., Keith-Spiegel, P.: Academic integrity as an institutional issue. Ethics Behav. **11**(3), 325–342 (2001)

28. Styron, J., Styron, R.A.: Student cheating and alternative web-based assessment. J. Coll. Teach. Learn. **7**(5), 37–42 (2010)

29. Manly, T.S., Leonard, L.N., Riemenschneider, C.K.: Academic integrity in the information age: virtues of respect and responsibility. J. Bus. Ethics **127**, 579–590 (2015)

30. O'Connell, J.: Networked participatory online learning design and challenges for academic integrity in higher education. Int. J. Educ. Integr. **12**(4). https://doi.org/10.1007/s40979-016-0009-7 (2016)

31. Jagiello-Rusilowski, A.: Drama for developing integrity in higher education. Palgrave Commun. https://doi.org/10.1057/palcomms.2017.29

32. Gamalel-Din, S.A.: Smart e-learning: a greater perspective; from the fourth to the fifth generation e-learning. Egypt. Inf. J. **11**, 39–48 (2010). https://doi.org/10.1016/j.eij.2010.06.006

33. Boehm, P.J., Justice, M., Weeks, S.: Promoting academic integrity in higher education. Community Coll. Enterp. **15**, 45–61 (2009)

34. Lofstrom, E., Trotman, T., Furnari, M., Shephard, K.: Who teaches academic integrity and how do they teach it? High. Educ. **69**, 435–448 (2015)

35. Supiano, B.: Faculty members at one more university push back at online programs. The Chronicle of Higher Education November 15, 2017

36. Mattes, M.: The private side of public higher education. https://tcf.org/content/report/private-side-public-higher-education/

Social Map Tool: Analysis of the Social Interactions of Elderly People in a Virtual Learning Environment

Leticia Rocha Machado[✉], Jozelina Mendes, Tássia Priscila Grande,
Larissa Justin, and Patricia Behar

Federal University of Rio Grande do Sul, Porto Alegre, Brazil
leticiarmachado@gmail.com,
jozelinasilvadasilva@gmail.com,
larissacjustin@gmail.com, tpri.fagundes@hotmail.com,
pbehar@terra.com.br

Abstract. The aim of this article is to analyze the social interactions of the elderly in a distance course in order to propose pedagogical strategies that can maximize the relationships in the virtual. The perspective of a long life for a large part of population poses new challenges in terms of technology, as public policies as individuals themselves. In this process discuss a permanent education during aging is fundamental. In this sense, there is a need to investigate the social interactions established in a virtual learning environment in Distance Learning, in order to build pedagogical strategies that enable the improvement of the quality of life of the elderly. Therefore, in order to reach the main objective of the research the methodology adopted was in a quanti-qualitative approach, composed of four stages: (1) Construction of theoretical framework; (2) Development of Distance Courses "Viv@EAD"; (3) Construction of Pedagogical Strategies for the social interactions of the elderly in Distance Learning; (4) Disclosure of results. Thirty elderly people participated in the research and at the blended course, over the 60 years. The results obtained showed a greater interaction when the elderly were instigated by the teachers of the course, being necessary the adoption of pedagogical strategies that make possible this engagement.

Keywords: Social map tool · Social interactions · Elderly people
Virtual learning environment

1 Introduction

With the popularization of the Internet, in recent years the use of online tools to offer virtual courses was started. Distance Learning (DE), as an education modality, could provide borderless learning for the elderly. Although some actions are already adopted for the older public, the data presented in a doctoral (PhD) thesis [1] point out the need to investigate and deepen the social relations established in this type of modality, since no research was found pointing out pedagogical strategies for the establishment of these interactions.

© Springer International Publishing AG, part of Springer Nature 2019
V. L. Uskov et al. (Eds.): KES SEEL-18 2018, SIST 99, pp. 165–174, 2019.
https://doi.org/10.1007/978-3-319-92363-5_15

The establishment of social interactions is fundamental in old age. Over time, social ties may diminish with the passing of friends, estrangement from the family, and intergenerational conflicts [2]. In this scenario, the virtual space available in Distance Learning can provided the necessary interactions at this time of life, allowing reflection on emotions and reminiscence, thus reducing the effects of aging [3]. The social interactions provided by the Internet are of interest to the elderly and should be more explored in courses aimed at the public in question.

Although there are already several experiences of specific courses for older people, there is still a lack of research on the impact and results of these courses, mainly investigations that consider the multiple facets of aging, considering that the elderly constitute a highly heterogeneous group. In this scenario, new learning environments are created, such as Distance Learning (DE).

Distance Learning is an education modality where the teaching and learning process occurs regardless of the time or space in which the subjects meet. Thus, with new paradigms, the need to investigate the educational use of digital technologies emerges, especially the social interactions carried out by the elderly.

Distance Learning can provide the elderly with the update that so many of them seek, as well as the possibility of the social interactions that are necessary at this stage of life. Unfortunately, in relation to the use of Distance Learning with the elderly, there are few experiences and researches in the field, both publications of experiences carried out in Brazil or abroad. However, it is understood that this modality of education can meet the diverse demands of the public with targeted content, time and specific methodology for the elders. Still, for this to occur, it is necessary to build specific pedagogical strategies for the elderly, since it allows the planning of effective learning spaces for the elderly, especially when considering the social aspects. There are many studies related to social interactions and/or social networks and the elderly, but few related to the use of digital technologies, especially in Distance Learning. Thus, this research had as main objective to analyze the social interactions of the elderly in a distance course in order to propose pedagogical strategies that can maximize the virtual relationships. For the development of the research, it was used the Social Map tool, available in the virtual learning environment ROODA (http://ead.ufrgs.br/rooda), that allows the presentation of sociograms of the social interactions carried out in the environment used. The data were collected in a course for the elderly called Viv@EAD - Living and learning in Distance Learning. This way, the objective is to highlight and promote the elderly as active subjects, combating prejudices and, consequently, creating conditions for a better quality of life.

2   Aging and Social Relations

The aging of the individual involves several aspects that should be seen individually and collectively. This process consists of a unique experience for each subject and can vary according to culture, sex, social class and the environment in which one lives, since human relations interfere directly in this process [4].

In this perspective, it is important to carry out investigations that broadly contemplate the facets of old age in order to discuss different ways of providing a better quality of life for the elderly, not only biologically, but psychologically and, especially, socially. Some researches in gerontology emphasize the importance of the elderly to have a social role, after all it is considered a significant point in the achievement of a successful aging, since it means (re) insertion in society.

Therefore, it is essential to think of spaces (educational or otherwise) and technological resources that increasingly enable the social interaction of this public who shows interest in the fields of digital technologies, as will be discussed below.

2.1 Social Interactions and Elderly: Educational Possibilities

The social relation is much investigated in the academic environment in different contexts and public. Some theories address the social issue, such as Blau's Theory of Exchanges [5], Rook's Theory of Equity [6] and Weiss's Theory of Social Provisions [7].

In education, Piaget presents the importance of the interactions of the subject with the object and this object may be other subjects. This interaction enables the construction of knowledge, it is through the actions that the construction is possible. As the author cites, "thought proceeds from action and a society is essentially a system of activities, whose elementary interactions consist in the proper sense of actions modifying one another according to certain laws of organization or equilibrium" [8]. In his publications, Freire [9] resumes the importance of the conscience of itself and the world. This dialectical movement, between the two consciousnesses, takes place through established relationships. It is in this relationship with the other that social interactions become fundamental.

Social interactions are fundamental in old age, but it is common for the social circle of the elderly to decrease due to the death of acquaintances, distancing from the family and a decrease in the involvement of social events [10].

The experiences in spaces that allow the creation of groups of coexistence usually strengthen the social role of the elderly, which tends to decrease after retirement [11]. In this context, it is considered that participation in activities, such as courses (face-to-face or distance), sports, volunteering, dance, etc., that can provide the composition of a social group is important to improve the quality of life of the [12].

In this perspective, the social interactions mediated by virtual tools are presented as an important option to meet this demand. Currently one of the main ways to interact socially is through digital social networks and the elderly have been increasingly becoming more and more present in these environments. Thus, when entering this context and conquering their space in the virtual world, the elderly feel socially active and included [13].

Another possibility is through educational actions, such as virtual courses in the Distance Learning modality. Machado and Behar [14] consider that "Distance Learning (DE) can become a rich space for the learning of older people, especially for the possibilities of social interaction and communication". The following will discuss possibilities for interaction through Distance Learning.

2.2 Distance Learning for Elderly

Distance Learning is a teaching modality where communication between those involved in the teaching and learning process occurs through the use of technologies, enabling the participants to be in different times and spaces [15].

Although they do not refer specifically to the elderly, census data from the Brazilian Association for Distance Learning (ABED) reveal that "distance-learning students tend to be older than students in face-to-face courses" [16]. This shows that this type of teaching has been gaining prominence in Brazil, especially among the older public.

Thus, from the perspective of increasing the use of technologies by the elderly, Distance Learning can become a teaching and learning possibility for this public. This modality of teaching can contribute to encourage lifelong learning since, by attending a distance course, the elderly can put their previous computer skills into practice [14].

There are many synchronous and asynchronous digital tools that can be used in DE, such as communication tools, task delivery and posting of content, among others, present in a Virtual Learning Environment (VLE). However, recently, other synchronous communication tools are also being used to aid education, such as WhatsApp, which is a mobile application used to exchange instant messages. Real-time digital communication tools provide a sense of instantaneousness, allowing greater social interaction [17].

In this context, using synchronous communication tools such as WhatsApp in DE can decrease the sense of distance and provide greater communication among participants in the teaching and learning process. This allows the elderly to expand their social interaction, contributing to their social and digital inclusion, enabling them to improve their knowledge in the use of technologies [18]. In this way it is important to analyze the social interactions in Distance Learning in order to propose pedagogical strategies that can promote these exchanges in the virtual.

3 Methodology

The present research was developed in a quantitative and qualitative approach. This format was chosen due to the viability of complementing the data, in order to assist in the understanding of the object of study. The target audience was 28 seniors aged 60 or over who participated in a distance learning course called "Viv@EAD - Living and Learning in Distance Learning" developed at the Federal University of Rio Grande do Sul - Brazil. In order to respond to the proposed goal, the present research was composed of four steps.

STEP 1 - Theoretical reference. At this stage, researches were carried out on the main topics covered in the research, mainly gerontological, social, educational and Distance Learning aspects.

STEP 2 - Development of ViV@EAD courses. Three workshops were planned and developed: Social relations in the virtual; Audiovisual Production and Healthy Aging. Each workshop was developed in a blended form and lasted for 1 month. For the social interactions and availability of the contents, the virtual learning environment ROODA (http://ead.ufrgs.br/rooda) was used. In addition to interaction features (Forum,

Messages, Logbook), it was also used a Library and Webfolio to post the activities performed by the participants. For the analysis of social interactions, the Social Map was also used, a feature that infers the social relations established in the environment (Fig. 1).

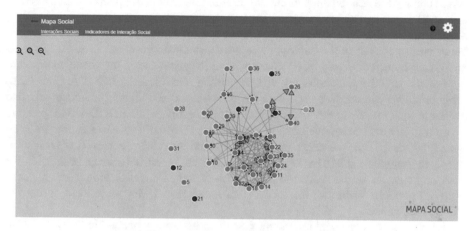

Fig. 1. Social map interface. Available at: http://ead.ufrgs.br/rooda.

In order to mediate the process and to propose the contents and activities, 3 teachers with a diverse background (pedagogy, language and computer science) participated. Also participating were 3 young monitors (aged 15–19) who were responsible on assisting the elderly in the difficulties related mainly to the handling of the technologies used in the course.

STEP 3 - Development of Pedagogical Strategies aimed at the social interactions of the elderly in Distance Learning. This stage was aimed at the development of pedagogical strategies, based on the analysis of the social interactions established by the elderly that can maximize a greater communication between the teachers of the elderly and/or the elderly. This construction was based on the notes made by the elderly in the functionalities of the VLE used (Step 2), as well as the sociograms generated by the Social Map and a questionnaire applied at the end of the course with open and closed questions.

STEP 4 - Dissemination of results. Depending on the data collected and its interpretation, the results will be the objects of publication of articles in periodicals, as well as presentations in national and international congresses.

The tools used for the data collection were: Participant observation, both in person and virtually; Sociograms generated by the Social Map functionality of VLE ROODA; Technological productions of the elderly in VLE ROODA; Questionnaire with open and closed questions. For the analysis of qualitative content, the steps suggested by Bardin [19] were used. The data collected and it's discussion will be presented below.

4   Results

The present research had the objective of analyzing the social interactions of the elderly in a distance course, called Viv@EAD, in order to propose pedagogical strategies that can maximize the relationships in the virtual. For that, 28 elderly people with a average age of 71.5 years participated, of which 86.4% were female. In relation to schooling, most of them have high school education and all of them live in Brazil. In order to better meet the needs of the elderly, these were separated into three classes (class A, B and C), each one composed of 10 students, both in person and in the virtual classroom.

For the social interactions the virtual learning environment ROODA (http://ead. ufrgs.br/rooda) was used. A page with the contents and activities of the class was also available for the elderly students: http://vivaead.weebly.com/. ROODA (Cooperative Learning Network) was developed at the Federal University of Rio Grande do Sul - Brazil, by the NUTED research group (http://nuted.ufrgs.br). Because it is an institutional environment, ROODA is constantly updated in order to keep pace with the emerging changes in the academic community. It has 26 functionalities, of which the Forum, Logbook, Webfolio, Contacts, Library and Social Map were used during the Viv@EAD course.

For the analysis of social interactions, sociograms were generated in the Social Map functionality. This tool has the objective to present graphs from the inference of messages exchanged in VLE ROODA.

For an in-depth analysis of the evolution of the interactions, sociograms were generated from the first week from the beginning of classes, followed by the second month of classes and the fourth month (final). The nodes represent the subjects who participated during the selected period. In orange color are the teachers, in the blue color the elderly students and in the green color the young monitors. In some situations it will appear in the color of wine those subjects who have escaped (who never entered VLE). The arrows that connect the nodes represent the interactions established, and the size of the tip represents the amount of messages exchanged between the subjects. Therefore, the following are the social interactions of class A in the first week and month 4 (Figs. 2 and 3). The names have been deleted to preserve the identity of the participants.

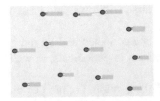

Fig. 2. Class A - first week.

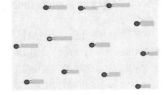

Fig. 3. Class A - 4th month (final).

Although there was a proposed activity so that students could interact in the first week, only one person performed this interaction through the forum. The other students did not exchange messages during the period (Fig. 2). In the second month the exchanges between the subjects intensified, with the interaction between the teacher-student and the elderly being more prevalent. Only 4 students exchanged messages with each other, using Contacts (messages) for this interaction.

In the last month the exchanges of messages from the teachers to the students intensified. Student C began a new contact with student J. The young female monitor did not contact and was not contacted by any student during the course period. Therefore, for class A, it is observed that the students interact little and the exchanges depart only from the teachers of the class (Fig. 3).

The following are the social interactions of class B elderly students (Figs. 4 and 5).

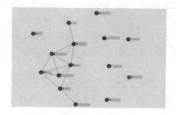

Fig. 4. Class B - first week.

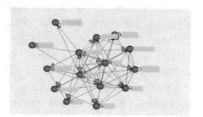

Fig. 5. Class B - 4th month (final).

However, class B had a higher interaction among students in the first week, with an average of 4 messages exchanged per student (Fig. 4). Most of the seniors used the Contacts for this communication. The elderly students intensified their exchanges, as can be seen in Fig. 5. It is also possible to observe in this group the involvement of the monitor with the participants, since she received messages by the contacts as well as sent them through the forum. The following are the sociograms of class C (Figs. 6 and 7).

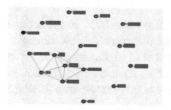

Fig. 6. Class C - first week.

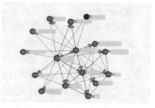

Fig. 7. Class C - 4th month (final).

Like class B (Fig. 5), this class also started the social exchanges in the first week, especially among some specific students (E, Na, Eli, T, A).

Interactions intensified, especially among teachers and students. The message exchanges between the students remained the same. An important data that can be analyzed is that the young monitor who had the task of assisting the elderly in the use of the technologies had not entered the environment until the selected period (it is highlighted in burgundy color).

With the data collected in the three classes in the different periods, it is possible to observe a greater interaction in class B compared to the other classes, mainly by the exchanges between the elderly students themselves. This data is mainly due to the time that this group has known each other (many since 2009) which allows a greater and faster approach among the members. Although the young monitor in class C (Figs. 5 and 6) did not enter the VLE, this did not directly reflect on the students' interactions, since, as it is possible to observe, they contacted the teachers to solve doubts and talk about another subjects.

Asked about the low interaction in VLE during the course, one of the elderly indicated that they prefer face-to-face classes since "In face-to-face social interactions we can always better clarify or solve some misunderstanding more directly, whereas in the virtual we can not always achieve that" (I21); "… always the face-to-face things captivate me more than the virtual ones because we have the eye in the eye that differentiates these relations" (I8). Despite these negative opinions about social interactions in the virtual, an elderly student stressed that "Chatting in classes (in person) and in the virtual has the same effect" (I27).

The collected data indicate that the elderly have a preference for face-to-face interactions, but they know the importance of the virtual ones, mainly because of the ease of communication. In relation to the social interactions carried out in VLE ROODA it is possible to observe the importance of the teachers' mediations to instigate the exchanges between the students. Therefore, it is possible to mention as pedagogical strategies that make it possible to maximize social interactions among the elderly: the providing of activities that instigate the use of communication tools in the VLE; to monitor the young monitors to carry out the interactions; activities that only allow communication between students; activities that facilitate the collaboration of materials among the elder students; group activities; current contents and related subjects with topics of public interest; use of the Social Map in the course of the process in order to follow up the interactions and thus modify the pedagogical practice if necessary.

It should be noted that these strategies will still be validated in 2017 in the second semester in a new course offered for the elderly public and that possibly new strategies will arise. Thus, it is observed with this investigation that it is important to carry out educational actions that foster social interactions in the virtual, since the participation of social groups means an interaction and creation of affective bonds that are important for the majority of the elderly.

5   Final Considerations

The technological advance is increasingly present in different areas, such as education, communication, health, entertainment, etc. Given this, keeping abreast of these innovations has become very important so that the individual can consider himself socially included. Thus, the number of elderly people who seek to know and deepen their knowledge about digital technologies, so that they can integrate into the contemporary panorama of society, has grown considerably in recent years. Therefore, it is necessary to always reflect on which pedagogical strategies can be adopted to cover all the needs of older people, mainly oriented to social interactions, whether face-to-face or virtual. In this sense, it is important to consider that this public has a different pace to follow and understand, especially in the current days, where the speed of information increases proportionally to the development of new technologies that lead them.

The collected data during the course of this study made it possible to analyze the social interactions of the elderly in different periods in a course offered in the Distance Learning modality. It was observed that those older people who a longer relationship in the classroom had a faster initiative to make exchanges in the virtual and had lasting social relationships during the course. Also the presence of the teachers instigating the interactions is of extreme importance for the establishment of social relations. Thus, some pedagogical strategies were developed and will be validated in the second half of 2017 in a new Distance Learning course for the elderly. Much still exists to research in this area, as well as new reflections and discussions about the role of education in this process of inclusion of older people.

References

1. Machado, L.: Construção de uma arquitetura pedagógica para cyberseniors: desvelando o potencial inclusivo da EAD. Tese (doutorado em Informática na Educação), UFRGS, Programa de Pós-Graduação em Informática na Educação, Porto Alegre (2013)
2. Castiglia, R.C., Pires, M.M., Boccardi, D.: Interação social do idoso frente a um programa de formação pessoal. RBCEH - Revista Brasileira de Ciências do Envelhecimento Humano (2006)
3. Pasqualotti, A.: Desenvolvimento dos aspectos sociais na velhice: experimentação de ambientes informatizados. In: Both, A., Barbosa, M.H.S., Benincá, C.R.S. (eds.) Envelhecimento humano: múltiplos olhares. UPF, Passo Fundo (2003)
4. Osório, A.R.: Os idosos na sociedade actual. In: Osório, A.R., Pinto, F.C. (orgs.) As pessoas idosas: contexto social e intervenção educativa. Instituto Piaget, Lisboa (2007)
5. Blau, P.: Exchange and Power in Social Life. Wiley, New York (1964)
6. Madden, M.: Older Adults and Social Media. http://www.pewinternet.org/2010/08/27/older-adults-and-social-media/
7. Rook, K., Pietromonaco, P.: Close relationships: ties that heal or ties that bind? In: Tones, W.H., Perlman, D. (eds.) Advances in Personal Relationships, vol. 1. JAI Press (1987)
8. Piaget, J.: Estudos sociológicos. Forense, Rio de Janeiro (1973)
9. Freire, P.: Pedagogia do oprimido. Paz e Terra, Rio de Janeiro (2002)
10. Scandolara, L.: A importância dos vínculos afetivos para o envelhecimento ativo. In: Terra, N., Bós, A., Castilhos, N. (Org.) Temas sobre envelhecimento ativo. EDIPUCRS, Porto Alegre (2013)
11. Rizolli, D., Surdi, A.C.: Percepção dos idosos sobre grupos de terceira idade. Revista Brasileira de Geriatria e Gerontologia 13(2), 225–233 (2010)
12. Mendes, J., et al.: Interações sociais de idosos: mapeamento de estratégias pedagógicas para Educação a Distância. In: Proceedings of the 12th Iberian Conference on Information Systems and Technologies (2017)
13. Machado et al.: Mapeamento De Competências Digitais: a inclusão social dos idosos. IN: ETD – Educ. Temat. Digit. Campinas, SP, vol. 18, no. 4, pp. 903–921, out./dez 2016
14. Machado, L.R., Behar, P.A.: Educação a Distância e Cybersêniors: um foco nas estratégias pedagógicas. IN: Educação & Realidade, Porto Alegre, vol. 40, no. 1, pp. 129–148, jan./mar 2015

15. Brasil. Decreto 9.027 de 25 de maio de 2017. Regulamenta o Art. 80 da Lei no 9.394, de 20 de dezembro de 1996. http://abed.org.br/arquivos/DECRETO_N_9.057_25_MAIO_2017_regulamentador_Educacao_Distancia.pdf
16. Censo EAD.BR: Relatório Analítico da Aprendizagem a Distância no Brasil (2014). http://www.abed.org.br/censoead2014/CensoEAD2014_portugues.pdf
17. Leite, L., Aguiar, M.: Tecnologia Educacional: das Práticas Tecnicistas à Cibercultura. In: Ramal, A., Santos, E. (Orgs.) Mídias e tecnologias na educação presencial e à distância. LTC, Rio de Janeiro (2016)
18. Lima, D., et al.: Avaliação da Primeira Experiência de Uso do Aplicativo WhatsApp por Usuários da Terceira Idade. In: II Conferência da Escola Regional de Sistemas de Informação do Rio de Janeiro. UFRJ, Rio de Janeiro (2015). http://www.lbd.dcc.ufmg.br/colecoes/ersi-rj/2015/011.pdf
19. Bardin, L.: Análise de Conteúdo. 4 ed. Edições 70, Lisboa (2010)

Smart Approaches in Facilitating Engineering Students to Learn Health Technology

Winson C. C. Lee[1,2](✉) and Gursel Alici[1,3]

[1] School of Mechanical, Materials, Mechatronic and Biomedical Engineering,
University of Wollongong, Wollongong, Australia
ccwlee@uow.edu.au
[2] Interdisciplinary Division of Biomedical Engineering, The Hong Kong
Polytechnic University, Kowloon, Hong Kong
[3] ARC Centre of Excellence for Electromaterials Science,
University of Wollongong, Wollongong, Australia

Abstract. Partly due to the aging population around the world, job markets for developing biomedical devices and technology have been rapidly expanding. Organizations are recruiting engineers of various disciplines to design and test biomedical devices. This arouses interest of some engineering students to take academic subjects to learn some health technology, while keeping their main disciplines in specific engineering areas like mechanical engineering. Engineers need to have good understanding of the human body and different sorts of diseases, before they can apply good engineering solutions to solve the medical problems. However, teaching engineering students without prior education in medical-related aspects could be challenging. This paper presents the use of e-learning and some other interactive teaching approaches to facilitate master-degree engineering students to learn biomedical devices and technology. The factors which may assist in success in multi-disciplinary learning and the possibility of implementing the similar in undergraduate courses are discussed.

Keywords: Engineering · Health technology · E-learning · Multi-disciplinary

1 Background

The aging population around the world is leading to an unprecedented need for medical devices and technology [1]. Australian researchers have found medical-device industry to be among one of the fastest growth sector, and called for measures to ensure supply of appropriately skilled graduates [2]. Globally, the market for medical device technologies has reached US $411.8 billion in 2013 and is expected to soar to over $530 billion by 2018 [3].

Biomedical engineers are trained to work in the industry of medical devices and technology. However, there were reports commenting that lack of medical content in biomedical engineering training had made engineers suffer from the "easier said than done" work in medical fields [4]. Some medical-device companies are recruiting engineers from other disciplines to do the job. For example, mechanical engineers are recruited to mechanically design and test orthopaedic implants. These workers could

© Springer International Publishing AG, part of Springer Nature 2019
V. L. Uskov et al. (Eds.): KES SEEL-18 2018, SIST 99, pp. 175–182, 2019.
https://doi.org/10.1007/978-3-319-92363-5_16

experience difficulty in communicating the medical needs with clinicians and patients [5]. Academia should identify ways to improve medical content in engineering curriculums and increase the flexibility of engineering students to take medical-related subjects to enable engineering graduates to have an option to work in this fast growing industry.

While Biomedical Engineering courses offer both medical-related (e.g. anatomy, physiology and subjects covering different diseases) and engineering (e.g. electronic and mechanical engineering) subjects, their student intake numbers are usually much smaller than traditional engineering courses. Some universities are trying to offer subjects which provide multi-disciplinary education allowing future engineers to sit together with future clinicians to learn the ways to solve health-related topics [6]. Engineers need to have good understanding of the human body and different sorts of diseases, before they can apply good engineering solutions to solve the medical problems. The challenges for engineers to learn medicine and related technology have been great.

This paper presents the use of e-learning and some other interactive teaching approaches to facilitate engineering students to learn biomedical devices and technology. The factors which may assist in success in multi-disciplinary learning and the possibilities of implementing the similar in different levels of education are discussed.

2   The Students and the Subject

Twenty-seven master-degree students were enrolled in the subject Rehabilitation Engineering. The first degrees of the students were mechanical (33%), electronic (30%), electrical (26%), civil (7%) engineering and others (4%). They were enrolled as either full-time (67%) or part-time (33%) taught master degree programs.

The subject Rehabilitation Engineering, which lasted 13 weeks over one academic session, was designed for taught (coursework) master-degree students. The subject aimed to provide students with knowledge on common neuro-musculoskeletal disorders as well as to review and simulate thinking over the use of engineering solutions to those medical problems. It consisted of 33 h (3 h/week) of combined lecture and tutorial as well 6 h of laboratory sessions. All teaching sessions were conducted at nights to allow some full-time workers to join in the class.

3   Teaching Approaches

3.1   Smart Learning Through the Internet

Learning Medical Disorders Online: Coping with the lack of in-depth knowledge in medicine of the students was one priority in this subject. While the students were provided with the names of common physical and sensory disorders as well as some general descriptions during the class, they were required to seek further information for the medical disorders. The students were requested to submit answers to some short questions in their assignments and quizzes, which tested their understanding of the medical terms.

Assessing Credibility of the Internet-Based Information: While the students relied on the internet for further reading on medical-related matters, discrepancies could exist among many different websites [7]. Students were reminded to assess the credibility of the materials they are reading. Factors important to credibility of online materials, which included author identification and qualification, presence of contact information, sponsorship by reputable organizations and citations to references [8], were discussed during the class. Feedbacks on assignments and quizzes also provided students with the sense of correctness and completeness of the medical information they have collected.

Encouraging Deep Learning: Students were also reminded to use higher-order thinking when reading the online materials. Such thinking involved evaluation and analysis of the information and creating new knowledge. In addition to assessing the credibility, students were advised to organize, test, compare/contrast different ideas and devise new thoughts. To facilitate students to employ higher-order thinking, there were involved in group projects (detailed in Sect. 3.4) requiring them to review and compare various approaches in applying technology to relieve a specific medical disorder and suggest ways of improvements. Feedbacks were provided to the students and students were required to refine their projects based on the suggestions.

3.2   Videos to Facilitate Understanding of Patients' Needs

Understanding the impairments caused by diseases or aging is very important before any engineering solutions can be devised. However, it is not easy to arrange transportation of different types of model patients with physical and sensory disabilities. Showing students videos could ease the problem. The teaching staff, with experience in real-life assessment of the disabilities, searched the internet and picked some appropriate videos to demonstrate to the students the effects of various diseases and degenerations on physical and sensory functions.

Two examples of using videos to teach medical problems are:

(1) Spasticity – a condition which involves involuntary contraction of muscles causing joint stiffness that affects many people with stroke. Videos obtained from Youtube^{TM} were shown to the students allowing them to learn how spasticity affects stroke patients and overall walking experience.
(2) Elderly gait – while textbooks show older adults have shorter stance time, step length and higher cadence compared to younger adults, students could have difficulty in grasping the characteristics of elderly gait. Students were shown some videos obtained from Youtube^{TM} demonstrating the common walking patterns of older adults.

3.3   Getting Students Involved in Smart Health Technology Projects

During the laboratory sessions, the teaching staff used their existing Smart Shoe project for older adults to stimulate thinking of students. The smart shoes detected any changes in walk stability of older adults during their long-distance walk and gave appropriate feedback based on their real-time detections. Students formed groups during the class to discuss the future directions of the projects. They were asked to report (1) what exact health problems the shoes could address, (2) why they thought these were significant, (3) how technology could be used to achieve their aims, and (4) the difficulties they foresaw. Students were encouraged to use their computers or smartphones to perform rapid search for information during the laboratory session.

3.4   Flexible Choice of Student Topics

One major assessment (40% of total subject scores) was a group-project assignment, which required the students to identify a neuro-musculoskeletal disorder of their interests, present existing technology used to alleviate the disorder, discuss the theory behind using appropriate medical and engineering terms, and suggest improvements. Students formed groups of 3–4 students. To ensure their topics align with the teaching aims of the subject, the students discussed their selected topics and discussed with the teaching staff during the class in week 3.

3.5   Feedback Through Online System

The students submitted their group-project assignments through Blackboard Online system (Blackboard Inc., Washington D.C., United States). They were required to summit an article showing the outline and some concrete information of their project in week 6. The teaching staff provided suggestions for improvements to individual group through the online system. Students were asked to revise their articles following the suggestions. Face-to-face discussion was arranged in week 11 to provide further suggestions to the revised article, before the final drafts were submitted in week 13. Instead of telling the students exactly what they should do, feedbacks from the teaching staff mostly involved questions which aimed to guide the students the possible ways of refining their projects. The Blackboard system generated email reminders of submission deadlines. Online forum was available which allowed the students and the staff to exchange ideas and information at any time.

4   Students' Performance and Feedback

Students' Performance. The students were assessed by Group projects (40%, also detailed in Sects. 3.4 and 3.5) as well as assignments and quizzes (60%, submitted by each student which assessed background knowledge of rehabilitation engineering). At the end, all students passed the subject. 11% of students received an overall rating of Excellent (Grade A or A+), and 67% received Good or Very Good ratings (Grade B or B+). Students generally performed well in group projects. They attained a mean (±standard deviation) score of 34.1 ± 4.1. Wider variability in scores (39.3 ± 13.2 out of 60) was observed in assignments and quizzes, which evaluated individual achievements.

Students' Feedback. In week 13, students were administered an online student feedback questionnaire by an administrative staff. Twenty-four students responded to the questionnaires (88.9% response rate). Figures 1, 2 and 3 show the findings regarding the subject, the teaching staff as well as overall learning experience. A majority (>70%) of students strongly agreed that (1) the staff member encouraged students to ask questions and discuss ideas in class, (2) the staff member encouraged to find information on their own and learn independently. Over 60% of students strongly agreed the teaching of the staff member has provided them with a valuable learning experience. The mean scores in each of the 11 assessment items reached top 10% of the entire Faculty of Engineering. In addition, a majority (63%) of students reported they spent on average 8–13 h per week on studying the subject. Most students (88%) thought the workload for this subject was appropriate, with the remaining 12% of students commenting the workload was too heavy.

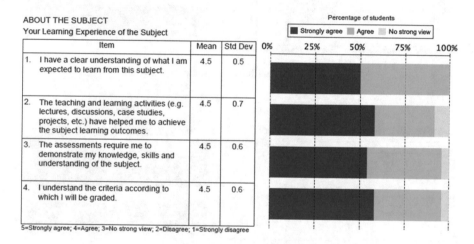

Fig. 1. Students' feedback about the subject.

Written feedback from students included:

"Give us many comments which are most helpful"

"Staff member's interaction with students has helped my learning"

"The assessments and feedback were most useful to my learning"

"Able to know the rehabilitation technologies better"

"Demonstrations on new prototypes are very helpful"

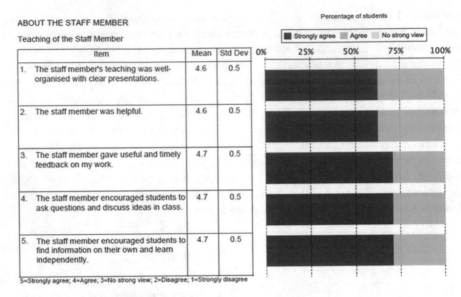

ABOUT THE STAFF MEMBER

Teaching of the Staff Member

Item	Mean	Std Dev
1. The staff member's teaching was well-organised with clear presentations.	4.6	0.5
2. The staff member was helpful.	4.6	0.5
3. The staff member gave useful and timely feedback on my work.	4.7	0.5
4. The staff member encouraged students to ask questions and discuss ideas in class.	4.7	0.5
5. The staff member encouraged students to find information on their own and learn independently.	4.7	0.5

5=Strongly agree; 4=Agree; 3=No strong view; 2=Disagree; 1=Strongly disagree

Fig. 2. Students' feedback about the teaching staff.

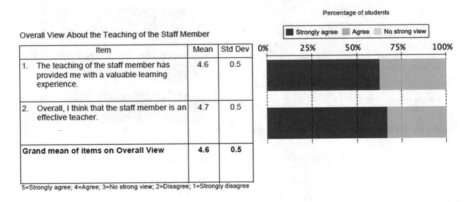

Overall View About the Teaching of the Staff Member

Item	Mean	Std Dev
1. The teaching of the staff member has provided me with a valuable learning experience.	4.6	0.5
2. Overall, I think that the staff member is an effective teacher.	4.7	0.5
Grand mean of items on Overall View	4.6	0.5

5=Strongly agree; 4=Agree; 3=No strong view; 2=Disagree; 1=Strongly disagree

Fig. 3. Students' overall impression.

5   Discussion

5.1   Technology-Supported and Interactive Teaching in Motivating Students to Learn

E-learning could facilitate active learning, which motivates students to learn [9]. A previous study also found active learning to be able to enhance students' performance in Science and Engineering [10]. This mode of learning involves students to engage in problems that involve higher order thinking tasks such as analysis, synthesis and evaluation [11]. To facilitate active learning, a balance should be struck between providing adequate support and challenge [12]. This could be supported by adoption of technology in teaching and learning. Use of websites to obtain medical knowledge and videos to further explain physical effects of the medical problems are good examples to this. Active learning could be further enhanced by providing students with some challenging questions, which require students to actively look for solutions using the internet and other sources. Allowing students to flexibly select a medical topic could align with students' interest, and a previous study has reported that students' academic motivation was dependent on their interest. Efforts should be made to help students to find the right information from the internet and other sources, and be able to interpret the collected information correctly. This can be supported by conducting academic assessments in the middle of the semester and providing feedback to students. Giving feedback to students is important, as a previous review stated that feedback was the most powerful single influence to student achievement [13]. And indeed many positive comments from students are related to the comments and feedback from the teaching staff. But resource constraints could lead to decline in quantity and quality of assessments and feedbacks. Providing feedback to group projects could save some manpower resource. In addition, online forum allows lecturers to provide feedback in a more casual way. To provide fair assessment, some assignments and quizzes should be arranged to test individual capability. Students have different learning characteristics, prior knowledge, personal needs and motivation. E-learning enables students to learn at their own pace, which is particularly useful in teaching a group of highly diversified students.

5.2   Implementation in Other Levels of Courses

While the subject was taught at master-degree level, it can be implemented in undergraduate level. Previous pilots have suggested the high feasibility of enabling undergraduate students with different academic background to address biomedical problems together [6, 14]. Engineering students require medical knowledge to enable them to use technology to solve health problems. However, their learning is different from medical students. While medical students have to systematically learn every aspect of a wide range of medical disorders, engineering students should learn the proper or pragmatic ways to obtain and interpret information when they are presented with a medical problem. Appropriate use of internet information to retrieve medical information could be one important aspect for engineering students to learn, and future research could identify better and easier ways of assessing correctness of internet-based information. To enable students to confront real-life problems, teaching staff may engage students to existing research projects. If they do not exist, students could be encouraged to review the existing engineering solutions to medical problems, identify limitations and propose improvements. The subject presented in this paper facilitated engineering students to learn about the use of technology in rehabilitation aids for people with physical disability. Similar teaching approaches may be applied to guide students to use engineering solutions to prevention, diagnosis and treatment of diseases.

6 Conclusion

This paper presents the use of e-learning and some other interactive teaching approaches to facilitate master-degree engineering students to learn biomedical devices and technology. Similar approaches may be applied to undergraduate courses to enable engineers to have an option to work in the fast growing medical industry.

References

1. Garçon, L., Khasnabis, C., Walker, L., Nakatani, Y., Lapitan, J., Borg, J., Ross, A., Velazquez, B.A.: Medical and assistive health technology: meeting the needs of aging populations. Gerontologist **56**(2), S293–S302 (2016)
2. Beddie, F., Creaser, M., Hargreaves, J., Ong, A.: Readiness to meet demand for skills: a study of five growth industries. Research report, National Centre for Vocational Education Research, Australia (2014)
3. BCC Research LLC. Medical devices: Technologies and global markets (2014)
4. Abdulhay, E., Khnouf, R., Haddad, S., Al-Bashir, A.: Improvement of medical content in the curriculum of biomedical engineering based on assessment of students outcomes. BMC Med. Educ. **17**(1), 129 (2017)
5. Linte, C.A.: Medicine through the eyes of an engineering: strengthening the engineer-physician collaborations. IEEE Eng. Med. Biol. Mag. **28**(5), 8–10 (2009)
6. Ludwig, P.M., Nagel, J.K., Lewis, E.J.: Student learning outcomes from a pilot medical innovations course with nursing, engineering, and biology undergraduate students. Int. J. STEM Educ. **4**, 33 (2017)
7. Davaris, M., Barnett, S., Abouassaly, R., Lawrentschuk, N.: Thoracic surgery information on the internet: a multilingual quality assessment. Interact J. Med. Res. **6**(1), e5 (2017)
8. Metzger, M.J.: Making sense of credibility on the web: models for evaluating online information and recommendations for future research. J. Am. Soc. Inf. Sci. Technol. **58**(13), 2078–2091 (2007)
9. Lumpkin, A., Achen, R.M., Dodd, R.K.: Student perceptions of active learning. Coll. Stud. J. **49**(1), 121–133 (2015)
10. Freeman, S., Eddy, S.L., McDonough, M., Smith, M.K., Okoroafor, N., Jordt, H., Wenderoth, M.P.: Active learning increases student performance in science, engineering, and mathematics. Proc. Natl. Acad. Sci. **111**(23), 8410–8415 (2014)
11. Sivan, A., Leung, R.W., Woon, C.C., Kember, D.: An implementation of active learning and its effect on the quality of student learning. Innov. Educ. Train. Int. **37**(4), 381–389 (2000)
12. Hunt, L., Chalmers, D., Macdonald, R.: Effective classroom teaching. In: Hunt, L., Chalmers, D. (eds.) University Teaching in Focus: A Learning-Centred Approach, pp. 21–37. ACER Press, Melbourne (2012)
13. Fraser, B.J.: Identifying the salient facets of a model of student learning: a synthesis of meta analyses. Int. J. Educ. Res. **11**(2), 187–212 (1987)
14. Cantillon-Murphy, P., McSweeney, J., Burgoyne, L., O'Tuathaigh, C., O'Flynn, S.: Addressing biomedical problems through interdisciplinary learning: a feasibility study. Int. J. Eng. Educ. **31**(1B), 1–10 (2015)

**Smart Education:
Case Studies and Research**

Enhancing Students' Involvement in the Process of Education Through Social Applications

Miloslava Cerna$^{(\boxtimes)}$ ⓘ and Anna Borkovcova

University of Hradec Kralove, Faculty of Informatics and Management,
Rokitanskeho 62, Hradec Kralove, Czech Republic
{miloslava.cerna, anna.borkovcova}@uhk.cz

Abstract. Findings presented in this paper contribute to the exploration of the evolving role of social applications in teaching/learning languages in engineering education. Utilization of internet sources, language websites and applications in the survey is approached from the perspective of students; it is based on 'Students' Language Needs Analysis'. The aim of the paper is to present selected findings relating to current language needs of students, which are relevant to the long-time explored issue on utilization of Web 2.0 in the university environment. The sub-goal is to propose ways of students' engagement into the studies reflecting their needs formulated in the 'Students' Language Needs Analysis'. Three quarters of students use the Internet for language study purposes. Website 'Help for English' is the most frequently used language portal, because it is a Czech portal and students know it from secondary school. Each participating respondent has his/her profile on the Facebook. But Facebook doesn't seem to be the proper place for systematic language use in the university setting because of many disturbing factors. A convenient solution might be a G + and its offer. A great deal of students has their profile on this net due to the Android but they use Facebook as a main communication channel. We believe that creation own classroom in this environment is a new challenge for both tutors and students.

Keywords: Education · Portals · Social application · Survey · Language

1 Introduction

Language educational portals, social applications, virtual learning space have been utilized in teaching/learning languages for many years. Utilization of the virtual world is not just a trendy technical innovation any more but it has become an inseparable part of the process of education going through all levels of the national educational system from the primary to tertiary education.

Findings presented in this paper contribute to the exploration of the evolving role of social applications in teaching/learning languages. Students' perspective dominates the research. In this contribution data are gained from selected relevant parts of the survey on students' language needs analysis. The survey with a follow-up discussion was run at the beginning of school semester with both part-time and full-time students of

© Springer International Publishing AG, part of Springer Nature 2019
V. L. Uskov et al. (Eds.): KES SEEL-18 2018, SIST 99, pp. 185–193, 2019.
https://doi.org/10.1007/978-3-319-92363-5_17

Information Management, Financial Management and Applied Informatics within Bachelor study programmes.

Utilization of social applications is a widely discussed issue among academics and teaching practitioners because this phenomenon has an undisputable drive in young people [1, 2]. Highly valued is the power of social media as a motivational tool [3, 4] to get students involved into the process of education.

Researchers from the Faculty of Informatics and Management have been involved in the issue of social applications within the frame of both national and university specific projects. This paper brings the latest data on utilization of applications for language purposes from the students' needs perspective.

2 State of Art

The issue of social applications in tertiary education is more often analyzed from the perspective of a tutor, [5] or facilitator [6]. Weller [7] discusses social applications from both perspectives from teachers' as well as students' perspective like, e.g. [8–10].

Utilization of internet sources, language websites and applications is in the survey approached from the perspective of students it is based on 'Students' Language Needs Analysis'.

2.1 The Aim of the Paper

The aim of the paper is to present selected findings relating to current language needs of students which are relevant to the long-time explored issue on utilization of Web 2.0 in the university environment. The sub-goal is to propose ways of students engagement into the studies reflecting their needs formulated in the 'Students' Language Needs Analysis'.

2.2 Procedure, Research Tool and Sample

The core of the paper is in the quantitative research. A questionnaire 'Students' Language Needs Analysis' was applied as a main research tool. Responses to the following selected questions (out of 10 questions) were analyzed:

- Q1 – Do you use the Internet for studying languages?
- Q2 – Which websites and applications do you use?
- Q3 – What functionalities should the application or website have to fit your language needs?
- Q5 – What would you like to practice most during "our" language classes?
- Q6 – What is the biggest trouble: Listening, reading, writing, grammar or just speaking?
- Q8 – What fits you most for studying languages?

Research sample consisted of 74 students: 19 part-time students studying subject Professional English I and 55 full-time students studying Professional English I and III, see Fig. 1. Abbreviation FT stands for full-time students and PT stands for part-time students. Two students were enrolled in both subjects. There were 45 males and 29 females.

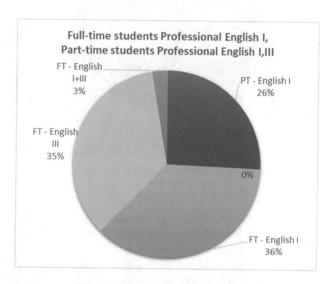

Fig. 1. Research sample and subject

During face-to face classes students were asked to fill in the 'Students' Language Needs Analysis', this questionnaire was placed into the English e-course in Blackboard management system. The e-course is systematically used as an enlarging space, communication tool, notice-board and repository of study materials. Out of 80 full-time students 55 students filled in and submitted their answers. As for part-time students the return of questionnaires was much lower due to the fact that there was only one face-to-face meeting, so that only one third of them submitted the questionnaire.

Selected inspiring findings from the questionnaires were discussed with full-time students during classes and useful links and recommended applications were placed into the e-course.

3   Findings

3.1   Do You Use Internet for Studying Languages?

Answers to the first question 'Do you use Internet for studying languages?' are visualized in the Fig. 2 Utilization of the Internet.

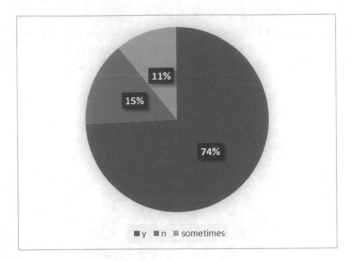

Fig. 2. Utilization of the Internet

Three quarters of students use the Internet for language study purposes. Only 15% stated that they do not use it for studying languages. Part-time students use the Internet even more then full-time students.

Not only part-time and full-time students differ in the use of the Internet when studying English language. There is also difference in the use of the Internet between men and women. Only two thirds of female students wrote that they use the Internet but there are four fifths of male students using the Internet in this sample.

3.2   Which Websites and Applications Do You Use?

Q2 – 'Which websites and applications do you use?' the second question was focused on students' experience with language websites and applications. Findings are illustrated in the Fig. 3. Used language websites and applications.

- Website 'Help for English' dominates. Students explained the popularity of the language portal especially by the fact that it is Czech portal, that they can find plentiful explanations in Czech, and that they had been familiar with it since their secondary school.
- Another highly represented web is BBC – Learning English with its abundant functionalities and links to main social nets.
- Wide popularity has gained the application 'Duolingo', in previous years those were mostly part-time students who used this application; currently the rate is nearly equal.
- Surprisingly, 'Youtube' as the most favorite social application in former surveys [2] was this time listed only rarely in this sample.

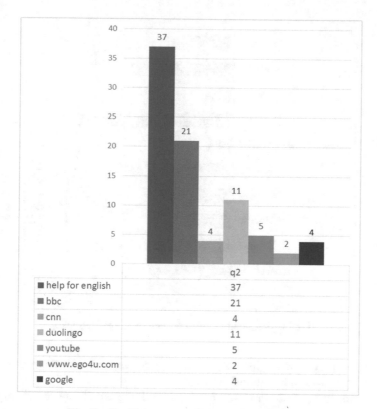

Fig. 3. Used language websites and applications

3.3 What Functionalities Should the Application or Website Have to Fit Your Language Needs?

Q3 was 'What functionalities should the application or website have to fit your language needs?' The third question was aiming at the applications themselves and students' specific requirements. Findings can be seen in the Fig. 4.

- Grammar seems to be a must; it has to be explained properly with given examples connected with real use. All respondents who mentioned the grammar called for exercises and tests:
  - These tests should be placed after each completed topic or level together with the assessment of good and bad results, in the ideal case also with the explanation and progress tracking.
  - Grammar should be categorized according to levels and according to topic.
  - Three students mentioned a need of a fun form.
- Easy navigation and use, easy search
- Accurate and reliable
- Everyday reminder
- Possibility to create own course and use also others' courses

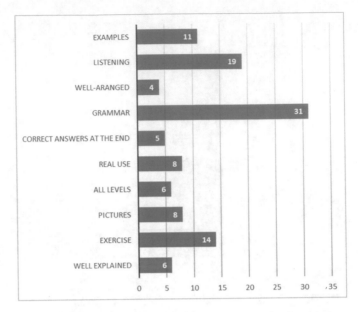

Fig. 4. Required features on language applications

- Everything purely in English (which doesn't correspond with their like in Czech website) but with hidden translation option.
- Kind of manuals, a summary of each topic with the possibility of printing
- Practicing pronunciation

3.4 What Would You like to Practice Most During "Our" Language Classes? What is the Biggest Trouble: Listening, Reading, Writing, Grammar or just Speaking?

Questions 5 and 6 from the Students' needs analysis were worked out into one graph. (Q5 – What would you like to practice most during "our" language classes? Q6 – What is the biggest trouble – Listening, reading, writing, grammar or just speaking? See Fig. 5.

Both questions were joined as the findings brought similar results.

Students require practicing speaking most of language skills. But they nearly ignore the need to develop reading skills. It is clear that face to face classes are the place for practicing oral language competences, students lack other possibilities.

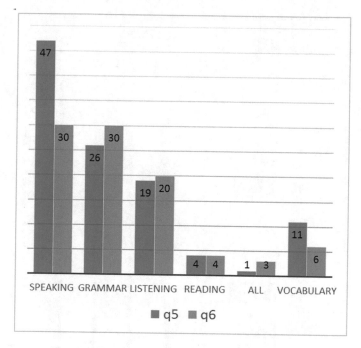

Fig. 5. What do the students need to practice most?

3.5   What Fits You Most for Studying Languages?

The final explored question is the question Q8 – What fits you most for studying languages? from the complex Students' language needs analysis. In this paper, it is the last discussed question. Results were placed into the Fig. 6.

- Movies are most frequently mentioned convenient language activity.
- Students value language websites and applications.

    The issues calling for discussion follow:

Is watching films really the most comfortable, entertaining and effective form of learning language?
What about face-to-face classes? Aren't they rather undervalued?

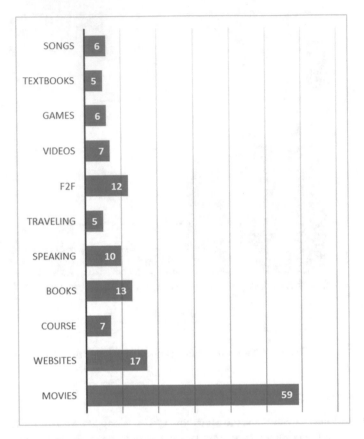

Fig. 6. Best activities for practicing language?

4   Conclusions

The goal of the paper was reached. An updated view of the current use of sources supporting practicing language in the case study in the small scale was provided.

The sub-goal was to propose ways of students' engagement into the studies reflecting their needs formulated in the 'Students' Language Needs Analysis'.

- Social nets can be used and it is desirable to use them also in the university setting. But they should be used as space for gaining information but not as an everyday study space. It is great place to learn about foreign events, discussion with foreign students and native speakers, sharing experience about foreign study stays and internships but not as a 'serious' study pace with tracking language progress, etc.
- Each participating respondent has his/her profile on the Facebook. But Facebook doesn't seem to be the proper place for systematic language use in the university setting as there are plentiful disturbing factors.
- A convenient solution might be a G + and its offer. A great deal of students has their profile on this net due to the Android but they use as a main communication

channel Facebook. We believe that creation own classroom in this environment is a new challenge for both tutors and students.
– The perceived important role of films could be used. Students could summarize the content, create questions, prepare a list of useful phrases, discuss cultural differences, etc.
– Application with a reminder for everyday study, enlargement of vocabulary and even test came from students themselves.

Acknowledgements. The paper is supported by the project SPEV 2017 at the Faculty of Informatics and Management of the University of Hradec Kralove, Czech Republic.

References

1. Cerna, M., Poulova, P.: Social software applications and their role in the process of education from the perspective of university students. In: Proceedings of the 11th European Conference on e-Learning, (ECEL 2012), Groningen, pp. 87–96 (2012)
2. Cerna, M.: Trends in acceptance of social software applications in higher education from the perspective of university students - case study. In: Proceedings of the 13th European Conference on e-Learning, ECEL 2014, Copenhagen (2014)
3. Silapachote, P., Srisuphab, A.: Gaining and maintaining student attention through competitive activities in cooperative learning, a well-received experience in an undergraduate introductory artificial intelligence course. In: Global Engineering Education Conference (EDUCON), Istanbul, 2014. IEEE (2014)
4. Jovanov, M., Gusev, M., Mihova, M.: The users' evaluation of an on-line collaborative activity for building ontology. In: Global Engineering Education Conference (EDUCON), Istanbul 2014. IEEE (2014)
5. Valtonen, T., et al.: Net generation at social software: Challenging assumptions, clarifying relationships and raising implications for learning. Int. J. Educ. Res. **49**, 210–219 (2010)
6. Tess, P.A.: The role of social media in higher education classes (real and virtual) – a literature review. Computers in Human Behavior (2013)
7. Weller, K. et. al.: Social Software in Academia: Three Studies on Users' Acceptance of Web 2.0 Services. Web Science Conference 2010, Raleigh (2010)
8. Klimova, B., Poulova, P.: Pedagogical principles of the implementation of social networks at schools. In: Lecture Notes in Computer Science (including subseries Lecture Notes in Artificial Intelligence and Lecture Notes in Bioinformatics), vol. 9584, LNCS, pp. 23–30. Springer, Cham (2016)
9. Poulova, P., Simonova, I.: Social networks supporting the higher education in the Czech Republic. In: Lecture Notes in Computer Science (including subseries Lecture Notes in Artificial Intelligence and Lecture Notes in Bioinformatics), vol. 9584, LNCS, pp. 67–76. Springer, Cham (2016)
10. Cerna, M., Poulova, P.: Social applications in engineering education. In: IEEE Global Engineering Education Conference, pp. 206–209. IEEE (2014)

Emerging Technologies, Social Computing and University Promotion

Petra Poulova[✉] and Blanka Klímová

University of Hradec Králové,
Rokitanského 62, 500 03 Hradec Králové, Czech Republic
{petra.poulova,blanka.klimova}@uhk.cz

Abstract. At the turn of the millennium, birth rates have declined very much in the Czech Republic as in other European countries. This demographic trend is currently reflected in the declining number of graduates and at the same time in the lower number of applicants for the university study. Universities must therefore intensively focus on marketing and the acquisition of high-quality students. In this demanding competition, emerging technologies and social computing play an important role in the whole process. The purpose of this article is to show how new media such as Facebook can contribute to the promotion of an institution of higher learning among future applicants for university study and in this way to attempt to acquire more potential candidates for this type of study.

Keywords: ICT · Internet · E-learning · Emerging technologies
Social computing · Marketing

1 Introduction

In the last decades of the 21st century, the penetration of information and communication technologies (ICT) is becoming increasingly frequent. A new informational society is being born. ICT is used not only in the economic, political and social fields but also in education.

The number of applicants for university studies is gradually declining. Universities compete among one another in order to acquire the highest-quality candidates for university study. The following numbers of students applying for the study at the Faculty of Informatics and Management University of Hradec Králové (FIM UHK) can demonstrate the whole situation. The number of applications exceeded six times the number of students who could have been enrolled in the first year of university studies in 2006 (3,608 applications – 693 students in the first year in October 2006), while in 2016 the number of applications was only three times higher than the number of students who could have been enrolled in the first year of university studies in 2016 (2,165 applications – 727 students in the first year in October 2016).

In this demanding competition, emerging technologies and social computing started to play an important role in the whole process. Especially, social network sites have

© Springer International Publishing AG, part of Springer Nature 2019
V. L. Uskov et al. (Eds.): KES SEEL-18 2018, SIST 99, pp. 194–203, 2019.
https://doi.org/10.1007/978-3-319-92363-5_18

become a widespread phenomenon nowadays since they can affect an enormous number of people through social networking. [1] And the most popular social network site among young people is Facebook. Research [2] indicates that 99% of young people use it. In addition, 87% of them exploit social networks every day. In fact, Facebook is a system which enables to create and maintain a list of mutually connected contacts, friends, who share their information and therefore such a social network site seems to be ideal for marketing and promotion.

Social network sites have distinctive characteristics which can help in distributing and promoting information. These include the following aspects [3]:

- user-based (social server is based on their users; without them there would be just pages with empty chats or calendar),
- community-driven (users are connected with each other by sharing the same or similar interests),
- interactive (besides the communication tools, the social networks contain different applications which can be exploited),
- relationships (users usually share mutual friendship), and
- emotion over content (the social networks provide their users with feeling of certainty, assurance that their friends are within an easy reach).

The purpose of this article is to show how new media such as Facebook can contribute to the promotion of an institution of higher learning among future applicants for university study and in this way to attempt to acquire more potential candidates for this type of study.

2   Smart and Emerging Technology in Education

As we look at the new trends in educational technology, we notice some common themes [4]:

- Individualism - much of the new technology is for one person and is customized to that person's preferences, i.e., to be used when convenient.
- Shared information - collaboration is the key, whether it is with increased reliance on social media and an opportunity for more openness.
- Hands on - even as education becomes more digitalized, there will be a need for hands-on learning.
- Flexibility - the learning process must be adapted to meet the needs of students.

Different authors list different technologies [5]. The most common are as follows:

- Wearable technology - smart watches and Google [4, 6]
- Mobile learning, Personalized Smartphone Apps [4, 6]
- BYOD - Bring your own device [4].
- 3D Printing [4, 6]

- Gamification, Virtual Reality [4, 6]
- Cloud computing [4]:
- Openness - open access journals, digital textbooks and other open content, open source software or MOOCs [4]
- Flipped, blended learning [4]
- Augmented Reality [6]
- Predictive Analytics [6]
- Digital Badges [6]

All these trends are gradually starting to be used at different levels of learning institutions and they affect not only education, but also the school management.

3   Role of Smart and Emerging Technologies and Social Computing in University Promotion

20 years ago, the faculty started to follow the information about the incentives which made the students to choose to study at FIM UHK.

3.1   Survey

The main method was a questionnaire. The questionnaire was anonymous and every new student was kindly asked to fill it in when he or she was being enrolled to FIM UHK.

The questionnaire was logically divided into several sections. The first one contains the questions concerning various communication channels. We were especially interested in the source where the student obtained the relevant information about our faculty and how many of them used our official www pages as an information source about the faculty.

The second part detected the ways students used for getting the results of his/her entrance exam.

The third part of the questionnaire was dedicated to student's possibilities of the Internet connection.

The fourth part focused on his/her experience in e-learning and the final part allowed the students to express any additional comments they might wish to share with the faculty.

3.2   Survey Response Rate

We prepared the questionnaire related to these issues already in 1996 and we managed to obtain a very high rate of responses in all the following years (consult Table 1).

Table 1. Survey response rate

	Enrolled students	Participating in survey	Return rate
1996/97	274	195	71%
1997/98	328	287	88%
1998/99	346	267	77%
1999/2000	292	238	82%
2000/01	384	311	81%
2001/02	338	321	95%
2002/03	430	350	81%
2003/04	615	498	81%
2004/05	654	525	80%
2005/06	637	517	81%
2006/07	693	500	72%
2007/08	866	625	72%
2008/09	895	526	59%
2009/10	809	476	59%
2010/11	934	551	59%
2011/12	932	644	69%
2012/13	1040	694	67%
2013/14	961	595	62%
2014/15	748	467	62%
2015/16	762	366	48%
2016/17	727	418	57%

3.3   The Efficiency of the Individual Communication Channels

The efficiency of the individual communication channels is depicted in Fig. 1. The efficiency rate represents the number of students that were addressed by the individual channels.

The information channels are as follows:

- Secondary School Teacher
- Educational Fair
- Educational consultant
- Others
- Phone call
- FIM Brochure
- Internet
- Facebook
- Open House
- FIM Students
- Press

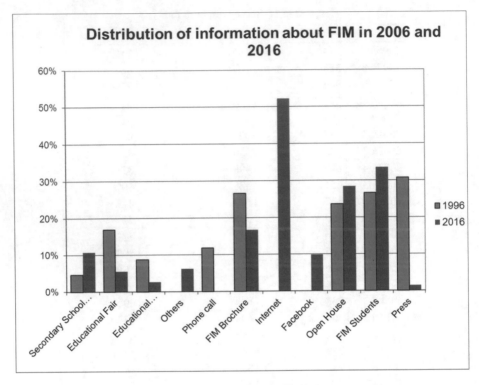

Fig. 1. Communication channels

Our former or the current students are the second the most important channel. We are delighted by our students' activities because their references are considered to be a very credible source of information for potential applicants. We support their interest by giving them a free access to our brochures, posters, and other printed materials that are distributed to secondary schools.

We can see that press and radio are very influential media in our case. We have developed very good relations with the relevant media representatives and that is why our messages are assumed to be a general publicity and we have not used any kind of paid advertisement yet. We do not save only our money, but it is also generally accepted that the audience pays much higher respect to publicity than to advertisements while the communicated message is the same one.

Our brochures and open-house day are another important channels that maintain their impact over 10%.

We can also observe preference for individual communication channels, which changed very much in the last several years. Press and radio was the most efficient communication channel before 2000 but now information leaderships was lost. (Consult Fig. 2). In addition, the role of the Internet significantly rose in the last three years. (See Fig. 3).

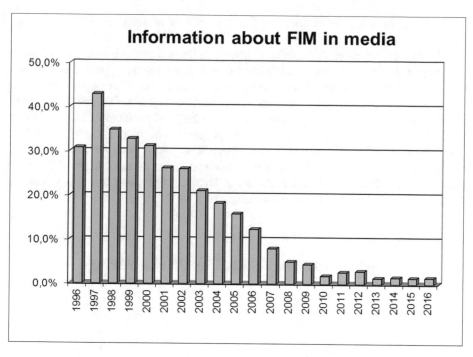

Fig. 2. Information about FIM in media

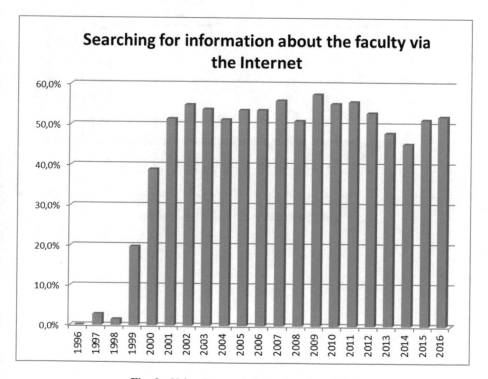

Fig. 3. Using Internet information about FIM

The role of other information channels levels off or it gradually decreases by frequent exploitation of the Internet.

The faculty started publish a huge amount of information about studying at FIM on the Internet in 1996. Any visitor of the faculty web pages can learn general information about the faculty, its offer of study programs, which are divided up to the level of template study program and syllabuses of a particular subject. Of course, there is also information about student's accommodation, boarding and the city of Hradec Králové.

If a visitor is interested in some study program, s/he can visit statistics of the entrance exams held in the previous year. This is a very good possibility to monitor the interest in each study program, results of previous students in dependence of academic record at secondary school and many other factors. The visitor can really measure his/her chance to be accepted to the faculty.

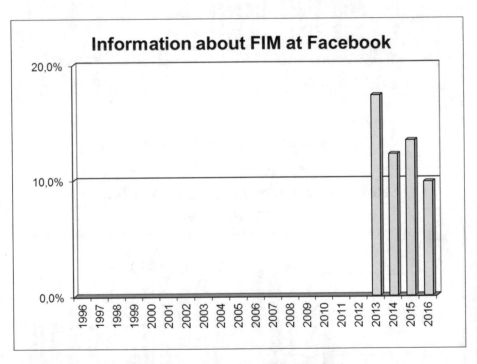

Fig. 4. Information about FIM at Facebook

In recent years the faculty started to use social network sites, especially Facebook, for its promotion and spread of information (Fig. 4).

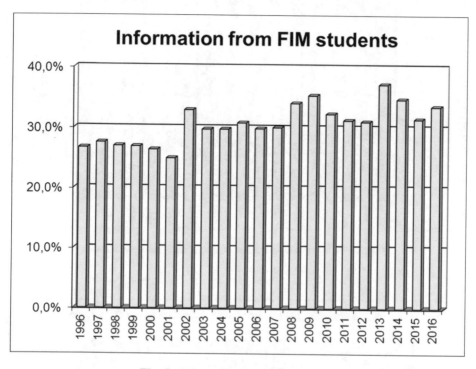

Fig. 5. Information from FIM students

Students seem to remain a stable information channel (Fig. 5).

Changes in student behavior are testified by the fact how many of applications in total the applicants place and whether they have been accepted to another university apart from FIM (Figs. 6 and 7).

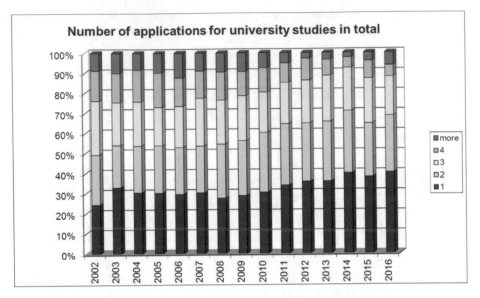

Fig. 6. Number of applications for university studies in total

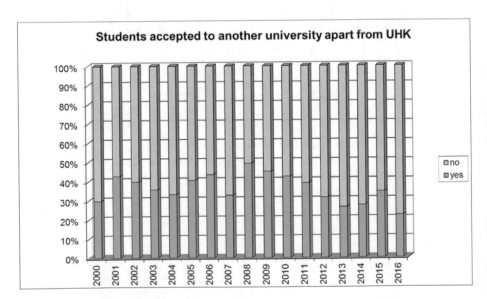

Fig. 7. Students accepted to another university apart from UHK

4 Conclusion

Overall, traditional media such as radio and newspaper have been declining in its popularity and information reach. Currently, new social media are on their rise. Universities are using social media platforms as new marketing tools, especially in attracting new potential students. And Facebook thanks to its widespread use appears to be the most prevalent and appropriate social medium for this purpose since the whole families and communities are on it. This analysis is applicable to other universities in the Czech Republic.

Acknowledgement. The paper is supported by the project SPEV (2018) at the Faculty of Informatics and Management of the University of Hradec Kralove, Czech Republic. In addition, the authors thank Jana Hamtilová for her help with the survey.

References

1. Klimova, B., Poulova, P., Ptackova, L.: Social network sites as a good support for study purposes. In: Uskov, V., Howlett, R.J., Jai, L.C. (eds.) Smart Innovation, Systems and Technologies, vol. 59, pp. 135–143. Springer (2016)
2. Klimova, B., Poulová, P.: Social networks and their potential for e-learning. Advances in Social and Behavioral Sciences. In: Proceedings of 2015 ACSS International Conference on the Social Sciences and Teaching Research (ACSS-SSTR 2015), Singapore, 6–7 December, 2015, pp. 3-8 (2015)
3. Dube, R.: Characteristics of social networks (2012). http://socialnetworking.lovetoknow.com/Characteristics_of_Social_Network
4. DeLoatch, P.: Emerging Education Technologies (2015). http://www.edudemic.com/10-emerging-education-technologies
5. Lewis, C.C., Fretwell, C.E., Ryan, J., Parham, J.B.: Faculty use of established and emerging technologies in higher education: a unified theory of acceptance and use of technology perspective. Int. J. High. Educ. **2**(2), 22–34 (2013)
6. Walsh, K.: 9 Educational Technologies That are Most Exciting Right Now (2016-17 Update) (2016). http://www.emergingedtech.com/2016/09/9-educational-technologies-that-are-most-exciting-right-now-2016-17-update/

Application of Smart-Education Technologies in the Institutions of the Russian System of Additional Education of Children

Svetlana A. Konovalova[1]([✉]), Nataliya I. Kashina[1],
Nataliya G. Tagiltseva[1], Svetlana V. Ward[2], Elvira M. Valeeva[3],
and Sergey I. Mokrousov[4]

[1] Ural State Pedagogical University, Yekaterinburg, Russia
konovsvetlana@mail.ru, musis52nt@mail.ru,
koranata@mail.ru
[2] Albany Creek State School, Albany Creek, Australia
morning_flower777@yahoo.com
[3] Southern Ural State University, Chelyabinsk, Russia
valelya@mail.ru
[4] Tyumen State University, Tyumen, Russia
yory67@mail.ru

Abstract. The article deals with the possibility of introducing the new toolkit in the segment of technologies Smart-education (computer (multimedia)) in the curricula of institutions of Russian supplementary education (children's music schools, children's art schools), the introduction of the educational process of the disciplines "Listening to Music" and "Musical Literature" presentation, video-clavier, electronic textbooks with audio and video sections, materials of sites related to musical art, multimedia programs for developing new children musical material in various formats). The teachers used this tool will successfully implement the personally oriented and activity-based methodological approaches in teaching: effectively organize the control of students' knowledge, build individual educational trajectories, successfully expand their musical experience, which determines the achievement of the goal of musical education - the formation of the musical culture of students as part of their common spiritual culture. The article also reveals the idea that the technologies of Smart-education are compatible with traditional methods of learning music. The effectiveness of the introduction of Smart-education technologies in the educational process of the disciplines "Listening to Music" and "Musical Literature" is confirmed by the qualitative and quantitative analysis of the experimental data using the capabilities of the statistical analysis apparatus (Pearson's criterion).

Keywords: Smart-education technologies · Children's music school
Children's art school · Additional education of children system
"Listening to Music" · "Music Literature"

© Springer International Publishing AG, part of Springer Nature 2019
V. L. Uskov et al. (Eds.): KES SEEL-18 2018, SIST 99, pp. 204–213, 2019.
https://doi.org/10.1007/978-3-319-92363-5_19

1 Introduction

At the present stage of the development of education, both in Russia and in other countries, there is an active implementation of a new toolkit that has a vast creative and pedagogical potential-Smart-education technologies that open up new opportunities for creating a "system of sociocultural innovations that open new opportunities for actors of education" [1]. They are based on the achievements of modern information and communication technologies and allow achieving new economic and social results.

In normative documents in the field of education and social development (the Federal Law "On Education in the Russian Federation", "The National Doctrine of Education in the Russian Federation", "The Concept of the Federal Targeted Program for the Development of Education for 2016–2020", the State Program of the Russian Federation "Information society (2011–2020), etc.) focuses attention on the need to ensure the development of the individual in the innovative conditions of education and upbringing; creation of programs implementing information technologies in education; training highly educated people and highly qualified professionals who are capable of professional growth and professional mobility in the context of informatization of the society and development of new science-intensive technologies.

All these requirements also apply to the system of Russian additional education for children, which is possible in the context of creative self-realization of the individual, due to a wide range of activities, the voluntary nature of work, a person-oriented approach to teaching and upbringing, a more free way of organizing the pedagogical process, based on the principles of identifying, supporting, developing the individual interests and abilities of each child who provides the maximum test conditions for the free choice of the sphere of creative activity in its various forms (performing vocal and instrumental arts) [2–4].

Today the researchers disclose didactic aspects of the problem of introducing Smart-education technologies in the system of Russian musical education (I.B. Gorbunova, P.L. Zhivaykin, I.M. Krasilnikov, S.P. Polozov, S.V. Puchkov, Y.N. Rags et al. They prove that Smart-education technologies are an important way of storing and broadcasting works of musical art; an element that promotes the formation of productive links between music theory and music practice; an element compatible and organically interacting with traditional methods of teaching music. These technologies do not replace the teacher, but contribute to increasing the effectiveness of the educational, creative and cultural-translational process carried out in the educational institutions of additional musical education.

In the educational-methodical laboratory at the Russian State Pedagogical University A.I. Herzen is developing the theoretical foundations and practices for the implementation of the concept of music and computer education in the training of music teachers. Smart-education technologies in this laboratory are considered as a new tool in the field of professional music education and creativity, a mean of teaching in general music education, a mean of rehabilitation for people with disabilities. In Russia, since 2003 is published the magazine "Music and Electronics", which reveals the problems associated with the selection of the necessary equipment for the teacher, the methodology for conducting lessons based on music and computer technologies, etc. Nevertheless, a number of issues remain "open".

The modern curricula of children's music schools and children's art schools include the disciplines "Listening to Music" and "Musical Literature". They are aimed at shaping students' knowledge about epochs and styles in history and art, the specificity of various musical instrumental and vocal genres, the development of interest and love to musical art, the development of musical perception, the formation of skills to work with music notation (clavier, score), etc. which enables the child to subsequently use this knowledge in his own creative work, to realize himself in one form or another of playing music.

Application of Smart-education technologies in the educational process will contribute to the effective organization of the learning process: for the expansion of their musical experience: they will allow the teacher to build individual educational trajectories, to implement personally oriented and activity-based methodological approaches in teaching students, because it is musical activity that is the "tool", the means of individual entering a person into culture [5].

In this regard, the development of ways to introduce Smart-education technologies into the content of the institutions of domestic supplementary education, in particular, to the educational process of the disciplines "Listening to Music" and "Musical Literature", are extremely relevant.

2 Theoretical Grounds for the Introduction of Smart-Education Technologies in Children's Music School and Art School

What are the technologies of Smart-education? In modern scientific literature is used the concept of "smart education", which is defined as an educational system that provides on the basis of the Internet interaction with the environment and the process of education and upbringing for citizens to acquire the necessary knowledge, skills, abilities and competences [6].

The concept of smart education (Smart-education) is associated with a number of concepts. The term "smart technology", which appeared in the scientific literature about 40 years ago in connection with the development of aerospace technologies, was later borrowed and disseminated by other branches of science [7]. The term "smart" implies to the property of a system or process that manifests itself in interaction with the environment and gives the system and/or process the ability to respond immediately to changes in the external environment; adaptation to changing conditions; independent development and self-control; effective achievement of the result. Smart, as a property, is most in demand in modern Russian social development and, in particular, in education, and is based on the achievements and development of information and communication technologies. The key to understanding of Smart-education is the wide availability of knowledge. Is flexible training in an interactive educational environment with the help of content from all over the world, which is freely available [8].

The analysis of numerous works on the introduction of Smart-education technologies into artistic and musical didactics allows us to draw a conclusion that they:

- have a complex impact on the perception of the user, including his emotional sphere (through the synthesis of visual, auditory and motor images in a single communication object, which contributes to the synergy of the recipient's perception and, accordingly, to a greater educational effect);
- create a sense of user presence at presentation art/music events (exhibitions, concerts, performances), in various buildings of the past (Gothic cathedrals, Orthodox churches, palaces of the Petrine era, etc.) for modeling the cultural environment, creating an artistic context for perceived music which leads to the activation of the associative thinking of the child, the finding of artistic parallels with the imagery of the musical work under study);
- facilitate the execution of a number of routine operations in the process of musical creativity (recording tracks and/or fragments, editing, arranging, listening to various sound variations, computer instrumentation), which contributes to its intensification;
- facilitate interaction (dialogue) of musical traditions, trends and genres, synthesis of various arts in culture (which contributes to the development of students' interest in the musical heritage of the past "for a deeper understanding of the students' ideas of the musical work and the search for ways of embodying their creative" I "in a specific artistic product" [9]);
- create an illusion of three-dimensional material objects, illusion of movement (allowing to enter the "space" of the artistic image of the work of art); - create representations using visual means of objects, the existence of which is impossible in objective reality, the reconstruction of the world of the past, the design of the world of the future (virtual modeling) [10]; expand the possibilities for independent search, analysis and subsequent generalization of necessary information.

The pedagogical potential of Smart-education technologies in the educational process "Listening to Music" and "Music Literature" is the following:

1. These technologies make it possible to optimize the process of teacher-student interaction due to the bright, imaginative teacher disclosing the material being studied, the implementation of the principles of clarity, activity and individualization that leads to more effective formation at students' knowledge of eras and styles in history and art, the specifics of various musical instrument and vocal genres. Even Jan Amos Kamensky in the XVII century pointed to the visibility as a "golden" rule of didactics. Later, at the beginning of the 20th century, K.D. Ushinsky wrote that the visualization corresponds to the nature of the child's thinking: "the child thinks with sounds, colors, forms, sensations in general ...". Modern authors (I.V. Kalinina, N.G. Tagiltseva, E.M. Torshilova) indicate the predisposition of children of different ages to the visual perception of modern culture. The reliance on clarity determines the features of the process of cognition, which always begins with a living contemplation.
2. Students have access to musical material audio and video format, the text information in the field of culture and art with the help of content from around the world,

is in the public domain (recording a performance of music by great musicians, conductors, vocalists, information about the life and work composers, etc.). This contributes to the development of their interest and love for the musical art, the development of their musical perception, the expansion of their personal musical experience, the experience of independent search and selection of necessary information.

3. In the process of studying a musical work, students get the opportunity to simultaneously perceive the music text and the sound of a musical work, which allows for an integral coverage of the musical form and artistic content, the formation of skills to work with the music text (clavier, score).

For teachers the use of Smart-education technologies to implement successfully a personality-oriented (individualization of instruction - the organization of the control of knowledge of students, the construction of individual educational trajectories, creating possibilities for their creative self-realization) ideational (training activities carried out using specialized software/technology). There sults of the development of cognitive interest of students, expansion of musical experience, which causes s goal of music education, declared in the current regulations in the field of music education of children with the formation of musical culture of the students as part of their common spiritual culture.

3   Experiment to Introduce Smart-Education Technologies in Children's Music School and Art School

3.1   Organizational Background of the Experiment

The main goal of the pedagogical experiment was to assess the effectiveness of introducing Smart-education technologies in the educational process of the two disciplines that are included in the curricula of modern Russian children's music schools and art schools: - the discipline "Listening to Music" (taught in the primary school), aimed at the formation of students in the culture of listening to music; - the discipline "Music Literature" (taught from the fourth to the graduating class), focused on the formation of musical thinking of students, the skills of perception and analysis of musical works, the acquisition of knowledge about the laws of musical form, the specificity of the musical language, expressive means of music.

Presented Smart-education technologies in the educational process of the children's music school № 2 named after. M.I. Glinka, the children's music school at the secondary school No. 32, the Children's School of Arts. ON. Rimsky-Korsakov, Yekaterinburg. 93 people participated in the experiment, of which the control and experimental groups were formed, as well as three teachers of music-theoretical disciplines. The experimental study included three stages: ascertaining, searching and control.

The aim of ascertaining stage of the experiment was to find out Smart-education technologies contributing to the expansion of student's musical experience. Criteria and indicators of the availability of musical experience musical experience (according to research by LV Scholar [11]), were: the level of general awareness of the music, there is an interest, certain tastes and preferences, treatment motivation child to this or that music. At this stage, diagnostic techniques were used L.V. Scholar "Meeting with the music in the classroom", "Music for the home library", "Musical program for friends" The results of this phase made it possible to identify the problem and studies confirm its relevance.

3.2   The Search Stage of the Experiment

The goal of the search phase was the introduction of Smart-education technologies into the educational process of the disciplines "Listening to Music" and "Music Literature". In the educational process of the mentioned above disciplines, the following programs were used to master the new musical material of the audio format: AIMP (developed by Russian programmers and supports all major audio formats, has the ability to convert various audio formats to any formats), PowerPoint, ProShowProducer, SonyVegas, WindowsMovieMaker, MediaPlayerClassic, WindowsMediaPlayer, iTunes (can be used to play music not only from computers, but also from smartphones based on iOs).

The study of new musical material (for example, scenic genres - operas, ballets, musicals performed by singers and world-class dancers) is better performed in video format using such programsas VLCMediaPlayer, MediaPlayerClassic, KMPlayer, WindowsMediaPlayer, Winamp.

In the educational process of the mentioned above disciplines were used a computer (multimedia) presentation, which is a small video film that contained graphic, textual and audiovisual information combined into a single structure. It was based on the storyline, script and navigation structure.

Multimedia presentation was a kind of "multimedia advertisement", that is, an artistic product that is characterized by integrity, expressiveness, the presence of a story and drama created by the teacher in the process of creativity with the help of multimedia tools to stimulate the interest of schoolchildren to a musical work or the composer's creativity [12]. Combining textual, graphic, animated, video and audio information in integrated form, multimedia technology enabled the teacher to work with text as well as visual design in the organization of educational material.

In the process of acquaintance of students with various operatic genres, it is possible to include students in a "live" choir into virtual opera action. For example, in the process of acquaintance of students with opera…. children were included in the concert by combining a video of the fragments of the opera (the best world theaters) and the performance of the children's choir of the music school (we also described this experience in [13]). This was achieved by alternately incorporating the performance of the children's choir and the video recording of the opera performance, which was projected onto a large screen.

The action was as follows (describe the performance of the boys' choir from the opera "The Queen of Spades" by PI Tchaikovsky). In the beginning, there was a video sequence with a fragment of the opera performance (Summer Garden, when the nurse, the foster-mothers and the governesses sang on the opera stage). Then the video recording stopped, and a real children's choir appeared on the stage - boys in the corresponding costumes. The children's choir performed a choral fragment ("We all gathered here in fear of the enemies of Russia ...". Then the video fragment of the opera was switched on again.

Thus, the audience could be present, on the one hand, in the real, on the other hand, virtual opera action, in which artists of the best opera theaters of the world took part, children participated in the choral collective of the children's art school.

Such a combined project gave children the opportunity to feel themselves in the role of an artist, formed an interest in opera and in the art of music in general. Modern computer technology, used in such an opera performance, was a mean of education and training.

In the educational process was also used such a modern mean as a video-clavier. Against the background of a video recording of an opera or other stage musical genre, a clavier recording of the work is entered into the video sequence, including the fixation of a musical notation (currently sung in the video fragment of the text). This greatly facilitated the learning by the children of the studied musical work, maked it more visible. These video-claviers are placed on the website of the pedagogical community of TA. Borovik - https://youtu.be/tO6304TI7v0. Such simultaneous listening and "tracing" the child text of the note "intact" is possible, thanks to information technology, when the sound is supplemented with a musical notation that appears on the computer screen simultaneously with the music of a particular work. Teacher used video-claviers: M.I. Glinka "Life for the Tsar", S.S. Prokofiev, the cantata "Alexander Nevsky."

Another direction in the use of Smart-education technologies in the disciplines "Listening to Music" and "Music Literature" was to teach students to search independently for necessary educational information. For this purpose the teacher can acquaint students with Internet sites containing information in the field of culture and art, as well as the music records in mp3; video recordings of opera and ballet performances, symphonic and chamber concerts, concerts of folk music groups; reproductions of paintings, consonant with the auditioned music; information about the life and work of composers, music figures, performers, etc. Let us cite some of them. It is: the Belcanto website (Electronic resource - http://www.belcanto.ru/), site "Classic-online" (Electronic resource - http://classic-online.ru), site "Intermezzo" (Electronic resource - http://www.aveclassics.net/board).

Students showed interest in certain musical works, they looked for their notes. In this case, they could use the sheet music, which was offered on the websites: http://www.notarhiv.ru/vokal.html; http://www.musicalarhive.ru/; http://xn–80aerctagto8a3d.xn–p1ai/ and http://notes.tarakanov.net/. They can listen and download music from the following sites: http://mp3ostrov.com/; http://gidmusic.net/classical and http://classic.chubrik.ru/ [14].

Another means of learning in the educational process of the disciplines "Listening to Music" and "Music Literature" was the use of electronic textbooks on music with audio-video sections:" Audiovisual Handbook on listening to music and studying opera and ballet genres" for 1st grade M.B. Kushnir. Also, an electronic tutorial was used in the Microsoft Power Point program on M.I. Shornikova "Musical literature. The development of Western European music", described by one of the authors of this article in [15].

3.3   Control Phase of the Experiment: Results of the Experiment

The purpose of the control phase of the experiment was to test the effectiveness of the introduction of Smart-education technologies in the educational process of the disciplines "Listening to Music" and "Music Literature." Comparative analysis of the initial and final sections showed a significant dynamic in the development of their musical experience.

The quantitative analysis of the experimental data was carried out with the help of the statistical analysis apparatus. The hypothesis tested was that the differences in the data in the initial and final sections are significant. Here the Pearson criterion was used. The theoretical value of the Pearson criterion is $X^2_T = 9.21$. The experimental value of the criterion X^2_e EG = 14.4. The experimental value of the criterion X^2_e of the group KG = 1.95. The result is: $(X^2_e$ EG = 14,4) > $(X^2_e$ CG = 1,95). The hypothesis is confirmed. Analysis of the theoretical and experimental values of the Pearson test for group A showed that $(X^2_e$ EG = 14.4) > $(X^2_T = 9.21)$, and therefore, between the levels of musical development of schoolchildren and the introduction of Smart-education technologies in this group, there is a dependence. A similar analysis of the data from the control group showed that there is no relationship between the levels of development of musical experience of schoolchildren and the stages of experimental work of communication $(X^2_e$ CG = 1,95) < $(X^2_T = 9,21)$. The analysis of the obtained data on EG and KG groups allowed to confirm the presence of the significance of the difference in these groups according to the investigated features.

4   Discussion

During the experiment the project authors faced the problem of some teachers' insufficient competence in Smart-education technologies. In order to solve this problem the upgrade qualification courses were organized on the basis of the Institute of Music and Art Education of the Ural State Pedagogical University. The authors of this article became the courses organizers. Successful development of students' creative self-realization and more complete revelation of children's unique potentials were noticed as one of the positive results of the experiment as well as some positive changes in the development of students' interest in classical music heritage.

5   Key Findings

The results of the introduction of technologies Smart-education (computer (multimedia) presentation, video-clavier, electronic textbooks with audio and video sections, materials of sites related to musical art, multimedia programs for the development of children of new musical material in various formats) into the educational process of disciplines "Listening to Music" and "Musical Literature" showed that these technologies influence the development of musical experience of schoolchildren - the level of general awareness of music (the assimilation of theoretical concepts, various phenomena of musical and social life, creative handwriting of various composers, peculiarities of the musical language, musical forms, genres, characteristic features of the sound of instruments of the symphony orchestra.

An experimental check of schoolchildren's educational achievements in mastering the content of the above-mentioned disciplines is carried out. The results of the experiment made it possible to draw a conclusion on the effectiveness of Smart-education technologies, which are one of the cultural forms of activity that oppose the tendency of the youth's interest in classical music heritage to decline. Their application demonstrates the mobility of domestic education, is a worthy response of educational structures to the social order of society.

6   Perspectives of Research

Conducted research reveals the perspectives of further Smart-education technologies implementation into educational process at children's music and art schools. Teachers' efforts can be aimed at the formation of senior students' abilities to make a computer (multimedia) presentation in the sphere of musical art and the formation of skills of students' self-sufficient activity using sheet music computer programs which have the resources to create timbre arrangements (instrumentations) of music pieces at home assignments.

References

1. Mokrousov, S.I., Kashina, N.I.: Internet resource of diagnostics and accompaniment of artistically gifted students. Pedagogical education in Russia, No. 11, pp. 72–76 (2017)
2. Kashina, N.I., Pavlov, D.N.: The problem of development of creative self-realization of college students of culture and arts in the process of musical composition activity. Innovative projects and programs in education, No. 3, pp. 11–15 (2016)
3. Konovalova, S.A.: Musical education and creativity. Modern problems of science and education, No. 1, pp. 58–59 (2006)
4. Konovalova, S.A.: Pedagogical model of the development of children's creative activity in music classes in conditions of additional education. Pedagogy of art, No. 4, pp. 106–112 (2011)
5. Kashina, N.I.: Peculiarities of the development of the musical culture of the Cossacks by modern students. Municipal formation: innovations and experiment, No. 1, pp. 57–61 (2016)
6. Dneprovskaya, N.V., Yankovskaya, E.A., Shevtsova, I.V.: Conceptual foundations of the concept of smart education. Open Education, No. 6, pp. 43–51 (2015)
7. Jeong, J.-S., Kim, M., Yoo, K.-H.: A Content oriented smart education system based on cloud computing. Int. J. Multimed. Ubiquitous Eng. **8**(6), 313–328 (2013). https://doi.org/10.14257/ijmue.2013.8.6.31
8. Zavrazhin, A.V.: SMART and new approaches in modern education. World of Education - Education in the World, No. 2, pp. 59–65 (2015)
9. Tagiltseva, N.G.: A polyartistic approach to the organization of the process of teaching music teachers in a pedagogical college. Pedagogical Education in Russia, No. 5, pp. 147–152 (2016)

10. Tagiltseva, N.G., Konovalova, S.A., Kashina, N.I., Valeeva, E.M., Ovsyannikova, O.A., Mokrousov, S.I.: Information technologies in musical and art education of children. In: Smart Innovation, Systems and Technologies, vol. 75, pp. 112–119 (2017)
11. Scholyar, L.V., Scholyar, V.A., Kritskaya, E.D. et al.: Music Education at School, Academy, 232 p. Moscow, Russia (2001). ISBN 5-7695-0443-9
12. Menshikova, N.A.: Reserves of multimedia technologies in the system of professional development of music teachers. Education and Science. Ural Branch of the Russian Academy of Education Review. Appendix, no. 2, pp. 33–37 (2006)
13. Tagiltseva, N.G., Russkikh, I.R.: Acquisition of junior schoolchildren to opera in the process of choral classes in the children's art school. Municipal formation: innovation and experiment, № 4, pp. 65–69 (2017)
14. Pacific Institute of Distance Education and Technology of the Far Eastern University, 208 p. (2004)
15. Tagiltseva, N.G., Prisyazhnaya, E.A.: Information Technologies in the Professional Activity of the Teacher of Additional Education. Municipal Formation: Innovation and Experiment, № 6, pp. 12–16 (2016)

Blended Learning 3.0: Getting Students on Board

Caron Eastgate Dann[(✉)]

School of Media, Film and Journalism, Monash University, Caulfield, Australia
caron.dann@monash.edu

Abstract. It is widely acknowledged that smart technologies offer exciting potential for university teaching and learning. In this brave new world, educators are urged to discard traditional teaching methods, such as the lecture, and replace them with student-centred modes that incorporate digital technology, self- and peer-based teaching and online content. In the communications and media studies discipline, many educators have been incorporating new technology for 10 years or more. However, we tend to use technology to enhance existing Material and methods: effectively, we are still using the lecture-tutorial teaching model of last century. We are approaching the third decade of the 21st century, yet research shows that real technology-driven change to teaching methods has been slow to develop. This is for many reasons, including institutional lag in providing access to cutting-edge software and networks. More often cited is educators' resistance, particularly in light of surveys in which students express dissatisfaction when blended or flipped classroom methods are introduced. I contend that the main reason for this dissatisfaction is about training: not of university staff, but of students, who we expect to automatically accept major changes to the way most have always been taught and assessed, without adequate explanation. Student training is part of what I see as the move to 'blended learning 3.0', in which we reassess what has worked and what has failed, on the way to revolutionising the way educators teach and students learn in 21st-century smart universities.

Keywords: Blended learning 3.0 · Student-centred learning · SoTL

1 Introduction

Students at the end of the second decade of this century are coming into university without the head-over-heels giddiness for technology that cohorts 10, 5 or even 3 years ago had. Part of the reason is that digital media with their apps, portable devices and live streaming possibilities have been normalised, losing the 'wow' factor. Gradually since the beginning of this millennium, and despite often clunky institutional systems, 'blended learning' has become the ideal at many universities in the West and beyond. 'Blended learning' is a vague term incorporating online systems in varying proportions for course material, assignment submission, readings, discussion forums and a/v resources. There have been strong warnings in the 21st century from education researchers, writers and designers, such as the US video game developer Marc Prensky (originator of the term 'digital native') and the US entrepreneur Salman Khan, that adding technology is not enough: educators must also change the way they teach. Prensky, Khan and many others advocate a 'flipped classroom', in which educators provide recorded lectures and other instructional parts of courses online, leaving face-to-face classes for mostly problem solving and peer-to-peer work in which students teach themselves with guidance from educators [1, 2].

Results have been mixed. End-of-semester student assessment surveys conducted after introduction of new methods are often disappointing. Given these discouraging results, a knee-jerk reaction from academics might be that flipped classroom and blended learning techniques don't work, because students don't engage well with them. However, in all this, I believe we have forgotten the main ingredient: the student (despite current best practice being described as 'student-centred learning'). While they are called 'digital natives', almost all my students remember what it was like before mobile media, and some remember what it was like not to have wifi, or even the internet at all, at home. Today's tertiary students had to learn how to use mobile media as it developed, just as their parents were doing. Importantly, almost all today's university students went to school in a non-flipped classroom environment: the teacher was the source of knowledge, textbooks were used, and online technology was generally a back-up to or expansion of traditional teaching methods. In short, new technology was seen as the entertainment, an adjunct to the 'real' teaching.

What I know about this area has been informed by 10 years of intensive teaching in tertiary-level communications. It seems to me that if we want to radically change the way we teach at university, we must also educate students about how and why we are making these changes and how they can best navigate them. For students, particularly international students, who have been used to a learning system in which the educator is the proverbial 'sage on the stage', it can be daunting to suddenly be expected to adapt to an atmosphere in which students teach themselves, learn from or even assess their peers, and in which the 'teacher' is not the central figure. Basically, we should follow the tenets of effective change-management: those involved need to be fully informed, aware of how the changes will affect them and of what they are expected to do. In this article, I make a case for providing such student training as part of our transition into what I call 'blended learning 3.0'. My argument is predicated on my interpretation of three stages in blended learning approaches to tertiary education:

Blended learning 1.0, late 1990 s-c2006. Realisation that new technology would be important in tertiary teaching methods, assessment and curriculum design.

Blended learning 2.0, c2007-present. Education designers reject traditional tertiary education modes, such as lectures, paper-based materials and teacher-centred learning, instead calling for educators to move to digital and online teaching resources and methods, and to student-centred learning. As mobile devices have become pervasive, there is increasing emphasis on innovation using new technology, including social media (after initial rejection of this in the 1.0 phase as inappropriate to tertiary student-teacher interaction).

Blended learning 3.0, 2018 and beyond. A reassessment of new technology and how it is used in tertiary education; reflection and critique on what works and what doesn't; a reappraisal of the different expectations of and dynamics between education designers, education deliverers and students, including investigation of new technology use in tertiary education.

2 Education Research: What Does It Tell Us About What Works, What Students Want, and What Educators Are Delivering?

Pedagogy emphasising student-centred learning is not new to education, particularly in schools, though less common in universities until recently. Many schools in the US experimented in the late 1960s and 1970s with 'Open Education' or 'Open Classroom'. The movement was labelled a failure by later educationalists, mainly because of its advocacy of classrooms free of walls, so that there would be several classes and many different group activities all proceeding at once [3], where noise levels were so high that no one could hear themselves think. However, Open Classroom education was much more than wall-free classes, and the philosophies behind it have influenced what today's education designers are prescribing. This is from 1975:

> *Information is not spoon-fed; wherever possible, it is built into a game that challenges the children to root it out for themselves...Today's children will have to perform in a world of exploding information. For them, it is more important to learn how to find the information than to commit a fraction of it to memory. Another effective resource available to every student is his fellow student... A wide variety of sophisticated learning aids not only encourages the children to teach themselves, but frees the teacher for more personal counselling... No longer limited to disseminating information from the head of the class, the teacher has become more like a doctor, who prescribes remedies for educational problems. She points the way, but the child teaches himself* [4].

Notice the similarity of 1970s teaching ideas to those of 21st-century student-centred teaching philosophy, such as this example from Singapore in 2012:

> *...[Today], knowledge is no more a monopoly among the teachers, because students can get knowledge from a myriad of sources, and hence, the role of the teacher today is facilitation. You know, that means facilitate students where they could get the right knowledge, how they could synthesise things, how they could discern the information that they get... —Professor Lee Sing Kong, Director, National Institute of Education, Singapore.*
> *They no longer become just a consumer of knowledge; they actually produce knowledge— Adrian Lim, Principal, Ngee Ann Secondary School* [5].

My point here is, although the ideas are not new, finally, we have the technology to realise their potential in allowing us to truly break away from the old 'teacher as god, student as follower' regime, in universities as well as schools.

Yet, we cannot simply present this new style of teaching at university as a *fait accompli*. While there is a perception that 'digital natives' are highly skilled at using technology and are able to adapt to it intuitively, the reality may be somewhat different, and we need to acknowledge the challenges both universities and students have in making productive use of new technology [6]. Education research in new media use started with examining pedagogical possibilities for academics to incorporate the new technology into practice. Some research has since moved on to examine what has worked and what hasn't and how students actually use technology. For a start, research shows that students mostly cite convenience over creativity [6]. Empirical research by Henderson, Selwyn & Aston involved surveying undergraduate students at two Australian public universities across all disciplines on their use of technology in learning in 2014. While ages varied from 17 to 66 across a sample of 1658 students, the mean age was 22.5 [6]. Results of this survey showed that students most valued their university's online system for keeping up with materials in classes, including dedicated weekly repositories such as Moodle and recorded lectures [6].

However, online course material does not necessarily represent innovation. Most online tertiary systems are merely the digital versions of pre-digital-era learning management systems that incorporated printed unit guides or outlines, book lists, class lists, and further reading materials. A recent study that surveyed 6139 academics in Italy concluded that 'Social Media use is still rather limited and restricted and that academics are not much inclined to integrate these devices into their practices for several reasons, such as cultural resistance, pedagogical issues or institutional constraints' [7]. The study contends that there needs to be more research on how social media is *actually* used by students and staff, rather than on its *potential* as a teaching aid. This is particularly so as 'the majority of studies published in the last decade mostly failed in establishing the technology's effectiveness at improving student learning' [7].

There is a growing chasm between 'the well-proven *potential* of technology-enabled learning and the less consistent *realities* of technology use within university teaching and learning' [6]. In many cases, digital systems represent *less* interactivity by students: digitised reading materials mean it is rarely necessary to visit a library; learning becomes a more passive at-home experience, with the replaying of lecture videos 'valued by some students more than any actual engagement' [6]. Students cited benefits as primarily those of logistics rather than learning: being able to study and view lectures anywhere, work on assignments until the last minute before they're due and not have to travel to campus to submit them, and having ready access to their online schedule so they didn't have to note classrooms and times [6].

Prensky contends that educators cannot simply use the old style of curriculum and plonk some new technology into the mix. In order for digital technology to work successfully in new blended learning classes, we have to rethink the content: 'Technology does not, and cannot, support the old pedagogy of telling/lecturing, except in the most minimal of ways, such as with pictures or videos. In fact, when teachers are using the old "telling" paradigm, adding technology, more often than not, gets in the way' [8]. Henderson, Selwyn & Aston's research shows that as educators, we need to promote innovation in tertiary education in ways that become engaging and meaningful for students. They found there were 'clear gaps between university students' actual uses of digital technology and the more abstracted rhetoric of "technology-enhanced-learning"...' [6]. Other studies have found that both academics and students perceive face-to-face teaching as more valuable than online teaching, and that social media 'are playing a marginal role in academic life' [7].

In coming to terms with the new educational landscape that digital technology offers, Prensky says we need to see education from the students' perspective, not from that of educators themselves. Prensky makes the point that this 'alternative' perspective allows us to adopt a bottom-up view that differs markedly from the perspectives 'nearly all educators, politicians, and parents are currently drawing' [8]. He is speaking primarily about K-12 teaching, but makes points that are useful to tertiary educators. On technology, he points out that although very important in the future of education, 'it does not dominate the vision; rather it supports it' [8]. He brings a view that asks for balance in education between technology and learning outcomes. Above all, technology should be used to assist teachers to 'find ways to create 21st century citizens (and workers) who parrot less and think more' [6].

A major challenge for the 21st-century tertiary educator is in fostering student engagement. With BYOD (bring your own device) learning environments, it is common that students interact with mobile media during class time. Sometimes this is to do with actual class content and activities; very often, not. Students might start texting their friends or checking their Facebook site because they are unengaged with the class. Despite the idea that engagement is crucial to higher education learning, there has been comparatively little empirical research on student engagement in university courses [9]. As noted above, there is more on how school education should or could change, but general principles are also applicable to university study.

3 Discussion: Blended Learning Fatigue? Ideals and Realities About What Students Want and What Teachers Can or Should Deliver

When I started lecturing in 2008, incorporating new technology was pretty much just maintaining a site online on the university's institutional system, which had a unit guide and the PowerPoint slides from lectures. Most units still had hard copies of readers, many required students to submit hard copies of essays and most had little interaction online. I didn't incorporate an online discussion forum assessment until 2015. This was a great success, and many colleagues engaged me in conversations about it. Many of us had forums by the start of 2016, and, by their survey comments, students generally enjoyed them, commenting that they were fresh, inclusive, and technology-focused. I noticed a sharp upturn in student engagement with the set readings, as the weekly online forum question was based on interpreting content from these readings.

Fast-forward to 2018, and students seem to have tired of forums. This is particularly so if the format requires student ingenuity, such as formulating questions for others students. A lecturer in late 2017 remarked to me that his students had 'hated the forums'. Obviously, the novelty value has worn off, and as so many units have these forums now, they've become commonplace, thus boring. Clearly, we need to rethink them. It's no longer good enough simply to add technology because it's there. As Prensky says, '[O]ur educators, in their push to get our classrooms and education up to date, too often add technology before the teachers know, pedagogically, what to do with it…[T]his often leads to the technology's becoming obsolete before it can ever add value' [8]. Prensky concludes that incorporating new technology in teaching is beneficial only when it is supported by 'a pedagogy of "partnering" ' [8]. The Australian media studies lecturer and researcher Andy Ruddock takes the idea of partnering further to make a case for 'an understanding of media education as a research practice involving partnerships between teachers and students', from first-year undergraduate level [10].

Many tertiary educators, particularly those involved in education design and enhancement, are declaring that 'the lecture is dead'. This, also, is not a new idea: Socrates criticised the lecture more than 2400 years ago, after all (although the Socratic method of debate in class—throwing questions to students to answer and discuss with the lecturer as leader—has been much criticised itself lately). Today, we are told by education experts, the class formerly known as the 'lecture' must be interactive, and that knowledge is ideally delivered online, recorded in small chunks rather than spoken by the lecturer in class. However, among practising lecturers, it is often a case of 'The lecture is dead; long live the lecture'. In theory, I agree that the stand-up lecture is no longer pedagogically sustainable. Yet in practice, it is difficult to get away from it completely. One reason it survives is because of large classes—some of 300 + students —particularly in undergraduate courses but also in some master's by coursework courses. Another is that there is an expectation by students that classes will be run by an expert who will tell them 'stuff'. In addition, many students have told me they equate online material with something that should be cheap or free—not the thousands of dollars in fees they pay for each unit they study.

Although I still give a 'lecture', my teaching practice has certainly changed since I started lecturing 10 years ago. I much prefer (but don't often get) flat-floor spaces with students grouped at tables, and I interact with students during lectures. I try to bring a personal and personable approach that enables them to feel comfortable to respond to questions, group exercises and online activity during 'lectures'. One notable factor is that my use of technology *during* lectures has produced: I play less a/v material in class, posting links for students online instead, because with mobile technology, they can easily check it out for themselves and we don't need to use valuable class time on it. The challenge is instilling in students the need to actually view, in their own time, material integral to in-class participation: watching a video online is considered an optional extra to a generation brought up with YouTube.

There are, at most sizable universities, short courses and workshops aimed at helping academics to incorporate new ideas and technologies in their teaching. Yet, this does not guarantee success: I know from conversations amongst educators that those who have made significant teaching changes using new technology are shocked if they receive unenthusiastic results in student surveys. In a 10-week foundation course for tertiary educators that I undertook in 2017, instructors agreed with course participants that flipped classrooms could lead to poor outcomes in student surveys. This is problematic, because—although it is widely agreed in academia that such surveys are flawed in several important ways—results are still used by the university as tools in assessing applications for promotion, in deciding teaching awards and as general teaching performance indicators, and also to identify educators who are judged as 'under-performers'. The safer option for academics, particularly those who are getting good survey results already, is to hang on to the old methods of teaching: a lecture with a 'sage on the stage' and a tutorial with stand-up student presentations and activities based on readings that (usually) few students have done.

One reason I see for students' low acceptance of new media in university teaching is the capabilities of students themselves. The majority of students are comfortable using their own devices and incorporating apps and other technologies they already know for social interaction, but most do not use technology in inventive or creative ways (11). New technology doesn't automatically bring creativity: instead, it provides tools students can use to help enhance their development as creative, innovative problem solvers, and this is where academics should come in [8].

It is clear that an evolution—if not a revolution—is taking place in how education is designed, delivered, and assessed, and that we cannot continue the old ways. With the internet, publicly available since 1993, have come profound changes to political, economic and social systems, and to workplace practices students will need to incorporate after graduation. The old higher education system, largely developed in the late 19th and early 20th centuries '... is no longer good at equipping graduates to succeed in an ever more complex and bewildering world' [11]. All disciplines are affected as digital natives—or, to use Prensky's more recent term, those with 'digital wisdom' [12] —become academics themselves and join university faculties.

4   Student-Centred Education in Blended Learning 3.0

Communications academics research many areas of their field, including audience and reception, institutional, political and economic analysis, celebrity and fan studies, and 'produser' culture. Few, it seems, are researching specifically what it means to be a communications educator in this changing technological era. Thus, my role in the next two years and beyond will be to begin to fill this gap in the scholarship of teaching and learning (SoTL) specifically in communications education at tertiary level.

The Australian academic Graeme Turner concludes his 2016 book *Re-inventing the Media* with a chapter titled 'Teaching the reinvented media', about the curriculum and content of what is being taught rather than about teaching methods, practices and evaluation [13]. However, implicit in Turner's analysis are lessons in the way media should be taught in future, and these can help inform teaching methodology and the scholarship of teaching and learning communications. Turner relates a conversation with a US academic, Michael Delli Carpini, in which the latter noted there were three standard ways of critically analysing media today: that despite technological changes, actually nothing much has changed; or, that everything has changed for the better; or, that everything has changed for the worse [13]. This could also be applied in scholarship to the analysis of pedagogy in media studies. In flipped and blended learning practices, then, we would look at what of the old is worth keeping; at what of the new is truly transformative; and at what of the new is detrimental to real learning, such as what happens to be fashionable at a particular time but will be short-lived. This fits with the idea that we need to think more deeply about how we are using digital technology in university education and what students really need [6]. Above all, we need to be thinking about how to equip students for long, productive lives in a work and economic landscape that is vastly different to that of the 20th century. In a multi-tasking, mobile, automated world, it is critical thinking, research and problem solving skills that will be most valued, while 'knowledge' takes care of itself.

Prensky contends that school students 'need an education that is far more connected and real than in the past—an education that gives them not only knowledge but also provides them with empowerment and agency' [12]. Applying this idea to a university context, and specifically to media studies, Ruddock contends that while *media literacy* has long been considered important to media studies (and beyond), media education in the 21st century should develop, as a priority, media *research* literacy [10]. To accomplish this requires the aforementioned partnership between students and teachers/researchers, in which students are active and interactive participants, not a row of faces looking toward the lecturer or tutor to tell them what's what. But how do we accomplish this partnership? This is a pertinent question, especially when it comes to logistics: most of our established universities, for example, are full of tiered 'lecture theatres' and crowded tutorial rooms with row-based seating facing the front. In contrast, the newest teaching spaces incorporate group desks that double as white-boards with technology charge-points; write-on walls and multiple interactive screens; spaces that enable instructors to more easily move round the room. However, these rooms are not yet the norm, and a problem is that you can design a unit one semester for a new-style teaching space, but next semester, you might be timetabled back to an old space. Thus, a challenge for academics is to try to incorporate innovative teaching while having to adjust depending on the teaching space offered.

Most importantly, we should be trying to foster student interest in—and demand for —new teaching and learning methods in smart interactive spaces. Until we make this happen, we will not be able to use educational technology advances to their potential. Currently, however, this is rare: '...digital technologies are clearly not "transforming" the nature of university teaching and learning, or even substantially disrupting the "student experience" ...[A]lternate contexts of teaching and learning need to be legitimized where alternate (perhaps more active, more participatory or more creative) uses of digital technology will be of genuine 'use' and 'help' [6].

Having recognised that students (as well as staff) need to be retrained to get the most of new blended-learning methods, I will start with a basic plan of what might work, and will back it with empirical research in the field over the next two years. My initial plan includes the following points:

(1) Explain at the first lecture what blended learning is, why it's being used in this unit, and how it differs from what students might have experienced at school or even at other universities. Explain why lectures are different to what they might expect from depictions they might have seen on media, particularly in films.
(2) For a successful blended learning environment, we must provide a schedule of what students are expected to access as a minimum each week, with clearly marked pre- and post-class activities.
(3) The student's learning journey could be an assessable component of the unit. For example, there could be a short reflective piece at the start of the unit about what areas the student needs to improve in order to get the best out of the unit, and a follow-up piece at the end reflecting on this progress.
(4) Explicit instructions should be given on how to best use social networking sites or other new technology required in the unit.

5   Conclusion

'Blended Learning 3.0' is the next step in technology-assisted university education, after moving through phases from the initial chaotic, bemused approach in incorporating new technology in our teaching (Blended Learning 1.0); to the love affair that championed all things digital, online and mobile (2.0); and finally, to today's quotidian atmosphere, in which online technologies have been normalised, but in the use of which there are varying degrees of proficiency and creativity.

This article has discussed the need for student training in blended learning methods. Students need to know what is expected of them, how to negotiate the new technology in terms of applying it to academic learning, and why changes are being made. It is not enough simply to post videos on a website and tell students to watch them and take notes. Most won't do either of those things, unless the watching of those videos is clearly integral to their progress in the unit, and unless they understand why they need to view them. The limitation of this contention is that you can't train away every challenge: that is, some things don't work because they don't work, even if students are adequately trained in the method. But direct instruction from educators as to how the unit is taught and why, what is expected from students and how to negotiate the technology will go a long way to improving student satisfaction with blended and flipped classroom techniques. My next step will be to test my hypothesis in a real classroom situation. My research will be grounded in my own practices: thus, as my teaching pedagogy evolves, it will inform my research.

One of the challenges to evaluating new teaching methods is how to measure their effectiveness. This is particularly so for educators who are already attaining high satisfaction rates in official student surveys. The surveys typically ask only about teaching, not about other aspects that affect a student's learning—a poky room without airconditioning and with outdated seating, for example, is not conducive to new teaching methods, but there is no mention of teaching spaces, at least on any survey I have seen. Because of this challenge, I propose to devise an alternative survey that I would ask students to complete in class time. The resulting data—both qualitative and quantitative—would then be analysed. In addition, I would plan to gather further in-depth qualitative data by holding focus groups and/or individual interviews with both students and staff. Naturally, this will depend on ethics clearance to proceed.

I hope to collaborate with educators within the communications discipline as well as further afield. From my experience, too often we are struggling alone to come up with solutions about how to incorporate new technology in teaching. There needs to be a stronger push for educators to share knowledge—of failures as well as successes. By the very nature of our discipline, we in the communications discipline imagined we were well up with using new technology in teaching and learning—but all the time, we were in danger of it passing us by and leaving us behind.

References

1. Prensky, M.: A Huge Leap for the Classroom: true peer-to-peer learning, enhanced by technology, Educational Technology, November–December 2011
2. Khan, S.: Let's teach for mastery—not test scores. TED talks (2016). https://www.youtube.com/watch?v=-MTRxRO5SRA. Accessed via YouTube 28 Nov 2017
3. Drummond, S.: "Open Schools" made noise in the '70 s; now they're just noisy', nprEd, National Public Radio, Inc. (US) (2017). https://www.npr.org/sections/ed/2017/03/27/520953343/open-schools-made-noise-in-the-70s-now-theyre-just-noisy. Accessed 23 Feb 2018
4. Historic Films Stock Footage Archive: Open Classroom—1975 (1975). https://www.youtube.com/watch?time_continue=290&v=z-DgYA_DKdA. Accessed via YouTube 12 Jan 2018
5. Education Everywhere Series: Singapore's 21st-Century Teaching Strategies. Edutopia Channel (2012). https://www.youtube.com/watch?v=M_pIK7ghGw4. Accessed via YouTube 1 Dec 2017
6. Henderson, M., Selwyn, N., Aston, R.: What works and why? Student perceptions of "useful" digital technology in university teaching and learning. Stud. High. Educ. **42**(8), 1567–1579 (2017)
7. Manca, S., Ranieri, M.: Facebook and the others. Potentials and obstacles of Social Media for teaching in higher education. Comput. Educ. **95**(April), 216–230 (2016)
8. Prensky, M.: From digital natives to digital wisdom: Hopeful essays for 21st century education. Corwin, Thousand Oaks (2012)
9. Manwaring, K., Larsen, R., Graham, C., Henrie, C., Halverson, L.: Investigating student engagement in blended learning settings using experience sampling and structural equation modeling. Internet High. Educ. **35**, 21–33 (2017)
10. Ruddock, A.: Exploring Media Research: Theories, Practice and Purpose. Sage, London (2017)
11. Davidson, C., Cook, C.: The New Education: How to Revolutionize the University to Prepare Students for a World in Flux. Basic Books, New York (2017)
12. Prensky, M.: Education to Better Their World: Unleashing the Power of 21st-Century Kids. Teachers College Press, New York (2016)
13. Turner, G.: Re-inventing the Media. Routledge, Milton Park (2016)

The Elderly in SMART Cities

Miloslava Cerna[✉], Petra Poulova, and Libuse Svobodova

Faculty of Informatics and Management, University of Hradec Kralove,
Rokitanskeho 62, Hradec Kralove, Czech Republic
{miloslava.cerna,petra.poulova,
libuse.svobodova}@uhk.cz

Abstract. With the concept of Smart cities, computer literacy as an inseparable part of everyday life becomes a kind of 'necessity'. This paper discusses requirements of the elderly people regarding skills in technologies and actual computer competency. The study, which is presented in this paper, brings findings from the research which was run by the authors on the local scene with a group of elderly people who attended computer courses within the frame of the 'Internet for the Senior Citizens' intergeneration project. The project is annually organized by municipality and primary schools to improve computer literacy in the senior attendees. Researchers applied quantitative and qualitative methods. Results relating to the respondents' history in gaining computer skills showed that self-education and assistance from family members represent the main role in this field, nearly one third of attendees marked attendance of computer courses. As for skills, which the elderly would like to gain or practice, work with photography and files dominated. Most attendees were familiar with E-mail; they found learning to use Skype communication application more beneficial. The aim of the paper is to identify computer skills that the seniors attending the free courses would like have or develop in order to be able to participate partially or fully in Smart City without feeling 'aside'. Research results were presented at the meeting of the directors and municipality representatives and are being incorporated into a next phase of the project with an updated design and content of the courses in a new learning environment.

Keywords: Elderly people · The Internet · Computer literacy
SMART city · Research

1 Introduction

A significant proportion of the population is made up of the elderly people this undisputable fact has to be taken into consideration when discussing, designing and applying the concept of SMART cities. This paper contributes to the issue of the elderly people and their skills in computer literacy, to their requirements regarding skills in technologies and actual computer competency. Preparedness of the elderly in the information and communication technology (ICT) area may foster their comfort in the changing environment and their wellbeing, not to feel aside the ongoing development.

© Springer International Publishing AG, part of Springer Nature 2019
V. L. Uskov et al. (Eds.): KES SEEL-18 2018, SIST 99, pp. 224–233, 2019.
https://doi.org/10.1007/978-3-319-92363-5_21

The European population is ageing, that is a plain fact validated by statistical authorities of individual countries. As for the Czech Republic, the main authority is a state Czech Statistical Office of the Czech Republic (CZSO) that collects and regularly updates data [1]. One of relevant and inspiring studies on the elderly people and utilization of technology is Casado ct al. [2] empirical study dealing with active ageing with a wide span of 55 to 94 years old participating respondents. Out of a great deal of definitions reflecting disciplines or perspectives of individual authors, a following definition of the 'Elderly people' was taken from the documents of the World Health Organization as the beginning of the senior age: "Most developed world countries have accepted the chronological age of 65 years as a definition of 'elderly' or older person. At the moment, there is no United Nations standard numerical criterion, but the UN agreed cutoff is 60+ years to refer to the older population." [3] The elderly people and utilization of technology is currently widely discussed topic with not always systematic solutions, sometimes coming ad-hoc. With the concept of Smart cities, the computer literacy as an inseparable part of everyday life becomes a kind of 'necessity'. An interesting finding is that only one fifth of people in the Czech Republic knows what the concept of Smart cities is. The concept of so-called smart solutions uses modern technology to improve the quality of life in cities. It focuses, among other things, on reducing energy consumption, sharing data for public purposes, or optimizing transport. Prague was among the first cities in our country that started to introduce the Smart city concept. Those who are familiar with this concept are mostly people from Prague, university graduates and people in the age category 25–34. The sample of the survey was representative because there were 1,500 respondents. [4] Smart city concept has to get into greater awareness; greater promotion and publicity is needed so that also senior generation could benefit from this concept. The issue has been part of the Faculty research interest for three years. The authors have participated in the research and published their findings; see [5–8]. The paper discusses researchers' findings from 'Internet for the Senior Citizens' municipality project. Currently they are involved in the updated version of content materials and learning environment of the new phase of the project.

2   Materials and Methods

Methodological background brings the aim of the study, description of the study procedure and literature review.

2.1   Internet for the Senior Citizens Project, the Aim, Phases and Sample of the Research

The study, which is presented in this paper, brings findings from the local scene; data were gained from the research that was conducted in the regional city reaching nearly 100,000 inhabitants where professional care to seniors is provided at various levels from home, via senior houses to courses on the state, municipality and charity basis. This study was run with a group of clderly people who attended free computer courses within the frame of the 'Internet for the Senior Citizens' project to improve their computer literacy. Municipality of the city of Hradec Kralove annually organizes this project. This commendable tradition was stablished eleven years ago. The awareness of this event got established among the city population due to systematic promotion on the websites of the city in the section designed for the elderly people [9] and in reports in local media [10].

The aim of the paper is to identify computer skills that the elderly people attending the municipality free courses would like have or develop in order to be able to participate partially or fully in Smart City without feeling 'aside'.

Phases of the research are following:

- Introduction to the topic
- Methodological background brings the aim of the study, description of the study procedure and literature review
- Next phase deals with the work 'in the field' with people involved into the long-term intergeneration project 'Internet for the Senior Citizens' which is organized by the Municipality of Hradec Kralove in twelve participating state primary schools
- Data from the project "Internet for the Senior Citizens" were collected from discussions, interviews, questionnaires and from the ceremony concluding the courses where the senior attendees were awarded with diplomas. Final phase focuses on findings gained from questionnaires.

Research sample consisted of people involved in the 'Internet for the Senior Citizens' project. Last year there were in total 173 senior citizens who accomplished a computer course in one of the twelve primary schools in the city. They successfully completed at least one of three kinds of courses: Internet for beginners, Internet for advanced computer users or Digital photography course.

Mixed qualitative and quantitative methods were applied in the research [11]. Semi-structured interviews with directors of involved schools were made and repeated discussions were carried out with senior participants, pupils and supervising teachers. A questionnaire was applied as one of key research tools. Four kinds of questionnaires were designed for four groups of participants: 'Student' questionnaire form for the senior participants, 'Lecturer' questionnaire form for to primary school pupils, 'Tutor' questionnaire form for the primary school teachers who ensured smooth run in each course, and finally 'Director' questionnaire form for the directors of schools. Researchers had an appointment with all directors and gained from them precious authentic material as well as from all tutors. Data from discussions and filled in questionnaires by two thirds of pupils – lecturers who assisted the seniors in the course were collected. Nearly half of senior citizens - 'students' provided researchers with the valuable response; 82 course attendees filled in the questionnaire. Researchers were given a warm support from the Deputy Mayor of the city who is in charge of education and social area in the municipality. She as a patron of the 'Internet for the Senior Citizens' project provided them with contacts, relevant materials and insightful look into the issue because she was at the design and repetitive run of the intergeneration project. The chronological age of the elderly people in this study was defined 65+ which correspondents to the definition given by WHO [2].

3   Literature Review

Literature review brings findings from professional and scientific literature, press, websites of selected studies and surveys, statistical offices (Czech Statistical Office and Eurostat) and other professional sources. It encompasses following areas: selected definitions to key expressions discussed in the paper, statistical data on utilization of telephones, computers and the Internet by the elderly people.

3.1   Smart Cities

SMART cities definitions were taken from the web portal The Centre for Cities which presents itself as the first port of call for UK and international decision makers seeking to understand and improve UK cities' economic performance. [12]. Three categories of definitions on SMART cities are given in this website: broad definitions, data-driven definitions and citizens-focused definitions. For the purposes of this study we have selected as the most appropriate 'citizens-focused definitions'.

"UK citizens tend to consider a smart city as clean, friendly and has good transport connections. Other words they associate with smart cities (although less frequently) include "technology", "connected", "internet" and "modern". [13] According to the Manchester Digital Development agency, "a 'smart city' means 'smart citizens' – where citizens have all the information they need to make informed choices about their lifestyle, work and travel options" [14].

3.2   Computer Literacy

Computer literacy ranks among not only useful but already essential skills. Definition on computer literacy was taken from the online dictionary Business dictionary [15]: Level of familiarity with the basic hardware and software (and now Internet) concepts that allows one to use personal computers for data entry, word processing, spread-sheets, and electronic communications. Next definition is from Techopedia. The IT Education Site defines Computer literate skill as follows: Computer literate is a term used to describe individuals who have the knowledge and skills to use a computer and other related technology. This term is usually used to describe the most basic knowledge and skills needed to operate soft-ware products such as an operating system, a software application, or an automated Web design tool [16].

3.3   Use of the Computer, Telephones and Internet by the Elderly – Statistics Data

When it comes to utilization of telephones, computers and the Internet by the elderly people in the Czech Republic, data are gained from the Czech Statistical Office [17]. Advanced technologies are in the Czech Republic commonly used for communication. In case of the analyzed elderly people category (65+) telephones are commonly used but computers and the Internet are used to much lesser extent. Following Tables 1 and 2 bring data showing in percentages the Internet use, households with home computers, households with the connection to the Internet.

Table 1. Individuals 65+ using the Internet in the Czech Republic; 2016 [17]

Total	Daily	On the mobile	At least once in life	In last 12 months	In last quartal
32,5%	18,7%	5,5%	43,2%	33,7%	32,5%

Table 2. Households in the Czech Republic with home computers and Internet connection [17]

	2012	2014	2016
65+ home computers	12,3%	23,8%	29,5%
65+ Internet connection	11,2%	22,7%	29,0%

A significant increase in households with home computers and households with the Internet connection can be seen. Roughly it rose three times. The important finding is that the Internet use is very close to the computer use in the elderly people category. 43% of seniors have used the Internet at least once and 33.7% in last three months. In the use of computers and their services to communicate with other people we still don´t achieve or even do not come close to the level of their utilization as is the average of the EU 28 [18].

4   Findings from the Research: Internet for Senior Citizens Courses with Focus on Computer Skills and Requirements

4.1   Selected Findings from the Discussions

Authors would like to highlight the social value and benefit of the project where the research was conducted. Deputy Mayor responsible for education and social area stated" … we can truly say that it is *meaningful activity* and brings a lot of good fruit. It is run in *the spirit of intergenerational dialogue*, but also *improves the quality of life* of our seniors. It contributes to their activation and prevents social isolation" [10].

One of the teachers who supervised the courses emphasized the *value of communication between two generations*; she enjoyed watching how 'her' pupils cope with their new role. Pupils from the lower secondary schools taught elderly people how to work with the computer in a ten-week course one hour weekly. She had sober expectations in the progress in computer literacy in the elderly people. She stated that there is no possible generalization in the technical achievements, because participants of the course differ. But several times she mentioned that *pupils change in perceiving the computer* – from predominantly useful thing for gaming to useful device of everyday use (organizing files, searching for information, e-mails. internet banking, etc.).

4.2   Findings from the Survey

The issue of computer literacy was explored from four perspectives. There are four groups of people who participate in the project who differ in cognitive and affective areas, they differ in knowledge and experience, and they differ in interest and expectations. Four kinds of modified questionnaires were created. Selected questions from the questionnaire fitting the scope of the conference were analyzed:

- Can you indicate how you have been educated in IT?
  - Self-study
  - My family helped me
  - My friends helped me
  - I have attended courses
  - This course is my first contact with my computer

Following set of computer skills is analyzed from three perspectives: respondents' actual computer skills, skills which they perceive as beneficial and contrary skills which they do not find important.

- Indicate what you would like to learn, what you already know and what is not important for you.
  - Use the Internet
  - Use Skype
  - Use social nets
  - Use e-mail
  - Work with photographs
  - Create and work with files

Results collected from the answers to the first question on seniors' history in gaining computer skills are visualized in Fig. 1.

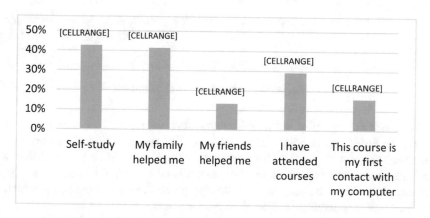

Fig. 1. Indicate how you have been educated in IT

'Self-education' and 'Help from family members' dominate, more than 40% of respondents marked this answer. The third most frequent answer is 'Attendance of computer courses' in nearly one third of attendees. *The crucial factor influencing the findings is that research sample consists of specific elderly people who are active and do not want to stagnate or stay aside. That is why it is not possible to generalize the results.*

The least frequent answer was 'Help from friends' or in other words friend's assistance. It can be assumed that friends will be in a similar situation with the similar background a have similar computer skills. Surprisingly there were 13 seniors from the sample, who had no experience with the computer and started to work with the computer for the first time in the course. The age of the seniors who got into the contact with a computer for the first time was from the age of 65 to the age of 81 with the even proportion of men and women. Gained findings might be used by the authorities responsible for the management of educational and care institutions when preparing and designing courses for the elderly people.

Figures 2, 3 and 4 illustrate results relating to the second question that focuses on the individual computer skills. Graph 2 shows what computer skills participants of the courses would like to learn.

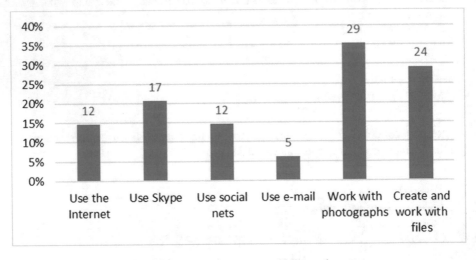

Fig. 2. Indicate what you would like to learn

Work with photography and files dominate. Astonishing 35% of participants would like to learn how to work with photographs and 30% would like to learn how to work with files. The least frequent answer relates to E-mail, it might be assumed that those who have some computer skills know how to use E-mail and find learning to use Skype more beneficial as can be seen in the following graph illustrating the skills, which the participants already have (based on their own answers).

The answers to the question 'Can you indicate what you already know' can be seen in the Fig. 3. 57 respondents that represents 70% of the sample claim that they know how to use the Internet and E-mail. More than 40% of course attendees can use the communication channel Skype. Only one fifth of them uses social networks that have become absolutely indispensable part of life for young people.

The fourth graph shows answers relating to computer skills, which the participants do not find important. It can be seen that only few attendees marked some of the skills, see Fig. 4. There were not many categories found unimportant by the elderly people.

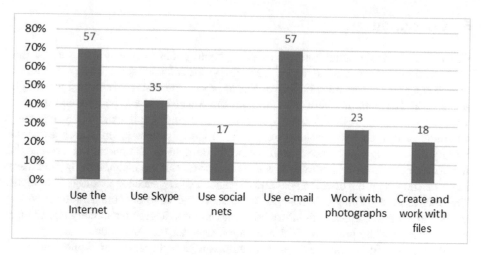

Fig. 3. Indicate what you already know

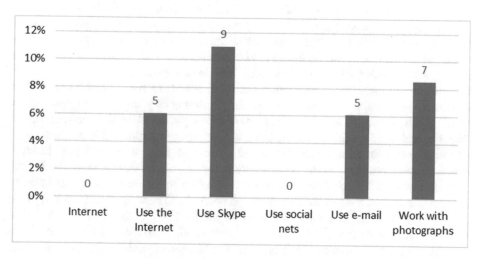

Fig. 4. Indicate what is not important for you

All participants consider the Internet and E-mail skills important. Only social nets exceeded the border of 10%, nine respondents marked that use of social nets is not important.

There were seven people who do not consider the skill 'Create and work with files' important. During the visits of the courses, it could be seen that some seniors have no or poor idea how to manage files. So the question arises: do the respondents understand what they are asked about? Do those who answered that file management is not important to them know what this work involves?

5   Conclusion

Creation of suitable conditions for elderly people could be the proper way to find solution for greater use of advanced technology, the internet, social networks, etc. As an inspiring and promising example of this approach the case study from Spain might serve. [19] The authors claim that although the Spanish elderly people are equipped with new technologies, they aren´t able to use them to full extent they use only fragment of offered options and possibilities. Elderly people in the Czech Republic widely use mobile phones but when it comes to computers, the Internet or even social network the percentage of active users is much lower. But the increase in Internet users in connection with the development and expansion of advanced technology is evident. The increased share might be caused by seniors who have become seniors in recent years and got into contact with information and communication technology e.g. in their work, which for the previous older generation was not very common.

Within the frame of the inter-generation project "Internet for Senior citizens" computer literacy in the Elderly people with focus on their requirements regarding skills in technologies and their actual computer competency was discussed. Results relating to the respondents' history in gaining computer skills showed that self-education and assistance from family members represented the main role in this field. Nearly one third of attendees marked attendance of computer courses. As for skills, which the elderly would like to gain or practice, work with photography and files was on the first place. Most attendees were familiar with E-mail; they found learning to use Skype communication application more beneficial. In the project itself, several missions combine like learning and socializing, meeting of young and old generations, gaining and sharing. Research results have been presented at the meeting of the directors and municipality representatives and are being incorporated into a new phase of the project with an updated design and content of the courses in a new learning space. Gained findings might be inspiration for the authorities responsible for the management of educational and care institutions when preparing courses for the elderly people. Data can be used in the economic sphere, in the marketing of ICT devices in the specific target group. The following work might focus an analyzing the gained data from another perspective, for example Joseph et al. [20] focused on a model for prioritization and prediction of impact of digital literacy training programmes and validation. Mustafa and Kar [21] focused on evaluating multi-dimensional risk for digital services in Smart cities.

Acknowledgement. The paper is supported by the project SPEV 2018 at the Faculty of Informatics and Management of the University of Hradec Kralove, Czech Republic. In addition, the authors thank Anna Borkovcova for her help with the project.

References

1. Age structure. https://www.czso.cz/csu/czso/animated-life-pyramids. Accessed 10 Jan 2018
2. Casado-Munoz, R., Lezcano, F., Rodriguez-Conde, M.J.: Active ageing and access to technology: an evolving empirical study. Comunicar **45**, 37–46 (2015)

3. Health statistics and information systems, Definition of an older or elderly person. http://www.who.int/healthinfo/survey/ageingdefnolder/en/. Accessed 10 Jan 2018

4. SMART city know only 1/5 of people in the CR. https://www.novinky.cz/internet-a-pc/459723-pojem-smart-city-zna-jen-petina-lidi-v-cesku.html. Accessed 10 Jan 2018

5. Hedvicakova, M., Svobodova, L.: Internet use by elderly people in the Czech Republic. In: Kar, A., et al. (eds.) Digital Nations – Smart Cities, Innovation, and Sustainability, I3E 2017, Lecture Notes in Computer Science, vol. 10595. Springer, Cham (2017)

6. Klimova, B.: Mobile phones and/or smartphones and their use in the management of dementia – findings from the research studies. In: Kar, A., et al. (eds.) Digital Nations – Smart Cities, Innovation, and Sustainability, I3E 2017, Lecture Notes in Computer Science, vol. 10595. Springer, Cham (2017)

7. Svobodova, L., Cerna, M.: Benefits and pitfalls in utilization of the Internet by elderly people. In: Kar, A., et al. (eds.) Digital Nations – Smart Cities, Innovation, and Sustainability, I3E 2017, Lecture Notes in Computer Science, vol. 10595. Springer, Cham (2017)

8. Svobodová, L., Hedvičáková, M.: The use of the social networks by elderly people in the Czech Republic and other countries V4. In: Kar, A., et al. (eds.) Digital Nations – Smart Cities, Innovation, and Sustainability, I3E 2017, Lecture Notes in Computer Science, vol. 10595. Springer, Cham (2017)

9. Autumn Internet courses for seniors - seniorhk.cz. http://www.hradeckralove.org/urad/internet-pro-seniory. Accessed 20 Jan 2018

10. Pupils in the role of real teachers. http://www.hradeckralove.org/noviny-a-novinky/zaci-v-roli-opravdovych-lektoru. Accessed 10 Jan 2018

11. Driscoll, D.L., Appiah-Yeboah, A., Salib., P Rupert, D.J.: Merging qualitative and quantitative data in mixed methods research: how to and why not. ecological and environmental anthropology (University of Georgia). http://digitalcommons.unl.edu/icwdmeea/18. Accessed 10 Jan 2018

12. Centre for cities. http://www.centreforcities.org/reader/smart-cities/what-is-a-smart-city/. Accessed 10 Jan 2018

13. Duckenfield, T.: What People Want From Their Cities, Connected Cities 2014. Steer Davies Gleave, London (2014)

14. MDDA. http://www.manchesterdda.com/smartcity/. Accessed 10 Jan 2018

15. Computer literacy – definition. http://www.businessdictionary.com/definition/computer-literacy.html. Accessed 10 Jan 2018

16. Definition - What does Computer Literate mean? https://www.techopedia.com/definition/23303/computer-literate. Accessed 10 Jan 2018

17. Czech statistical office. Information Society in Figures. https://www.czso.cz/csu/czso/information_society_in_figures. Accessed 10 Jan 2018

18. Eurostat. http://ec.europa.eu/eurostat/statisticsexplained/index.php/Digital_economy_and_society_statistics_-_households_and_individuals. Accessed 10 Jan 2018

19. Llorente-Barroso, C., Vinaras-Abad, M., Sancher-Valle, M.: Internet and the elderly: enhancing active ageing. Comunicar **45**, 29–36 (2015)

20. Joseph, N., Kar, A.K., Ilavarasan, P.V.: A model for prioritization and prediction of impact of digital literacy training programmes and validation. In: Kar, A., et al. (eds.) Digital Nations – Smart Cities, Innovation, and Sustainability, I3E 2017, Lecture Notes in Computer Science, vol. 10595. Springer, Cham (2017)

21. Mustafa, S.Z., Kar, A.K.: Evaluating multi-dimensional risk for digital services in smart cities. In: Kar, A., et al. (eds.) Digital Nations – Smart Cities, Innovation, and Sustainability, I3E 2017, Lecture Notes in Computer Science, vol. 10595. Springer, Cham (2017)

A Framework for Planning and Control
of the Education Organization

Rudy Oude Vrielink[1(✉)], Verin Nijhuis-Boer[2], Carien van Horne[2],
Erwin Hans[1], and Jos van Hillegersberg[1]

[1] Department of Industrial Engineering and Business Information Management,
University of Twente, Enschede, The Netherlands
r.a.oudevrielink@utwente.nl
[2] Saxion University of Applied Science, Enschede, The Netherlands

Abstract. Education logistics concerns the integral planning and control (P&C) of education related activities in education organizations, encompassing activities such as scheduling, facility management, staffing, performance management, curriculum development, and ICT management. It has a significant impact on the overall effectiveness and efficiency of the organization. The use of P&C needs to transcend the overtone of finance and aim for the much broader meaning of integral coordination or the organization. This paper introduces a new framework for education organizations to design, optimize, and govern their education logistics and its developments. It introduces hierarchical layers of P&C in a similar way as existing theoretical P&C frameworks for other sectors such as industry and healthcare, and distinguishes different process areas within these layers. We demonstrate that the current dominant way of P&C in education is falling short in numerous aspects, thereby impeding both innovation and the effective and efficient implementation of emerging educational paradigms. Our framework for smart education organizations counters current P&C deficiencies by closing existing coordination gaps between directive-, primary-, and support processes. Currently, these process areas in higher education organizations are often functionally dispersed through lack of horizontal alignment. Our framework defines and integrates levels of control for different process areas, thereby integrating and aligning decision making in the organization.

Keywords: Education logistics · Governance · Planning and control

1 Introduction

The educational landscape is rapidly changing, through new education paradigms like problem, project- or student-driven learning, and emerging ICT innovations in- and outside the classroom [1, 2]. Gaps in coordination in education organizations through lack of coordination between distinctive process areas, show that support services find it hard to keep up with the changes in education, which frustrates innovation and leads to inefficiencies. This paper focuses on the governance of (higher) education organizations (HEIs), through education planning and control, which is also referred to as education logistics. Education logistics is defined as "the sum of processes, systems

© Springer International Publishing AG, part of Springer Nature 2019
V. L. Uskov et al. (Eds.): KES SEEL-18 2018, SIST 99, pp. 234–245, 2019.
https://doi.org/10.1007/978-3-319-92363-5_22

and information flows that facilitate education at universities and other education organizations in a streamlined way" [3]. Beyond the efficient and effective planning of educational events in classrooms, it is also the timely creation of rules and regulations, design of tests and exams, drawing the curriculum, planning of the staff, matching the students to the courses, evaluating, and informing management about the process flows. Education logistics strives to control the entire education organization, and is ideally a chain of collaborative education processes with integral ICT support in the organization and in relation with its chain partners. In education, there is a growing attention for integration of all aspects of support to improve alignment with the primary processes [4], aiming at increasing customer satisfaction and with the intention of a better allocation of people and resources. Therefore, it is necessary to integrally plan and control education processes. Unfortunately, reality shows that this is not always the case [5].

Sir Ken Robinson [6] described in "education 2.0", the paradigm shift he sees in education. Education is shifting towards more student-centered approaches, enabling students to study in their own pace, in their own time, and with their own learning curve. Students are more and more seen as customers, and the organization has to adapt to the educational needs of each customer. HEIs increasingly experiment and apply new didactical concepts, striving not only for more attractive education, but also to improve on success rates, rankings and other effectiveness of education. New didactical concepts such as blended learning and 'flip the classroom' are used by lecturers to increase attractiveness of teaching, and also aim to get the most out of the fact that they have the students with them in the classroom. Education 2.0 has also asserted the importance of inter- and multi-disciplinary skills. Particularly in higher education programs, students are educated to become professional specialists who will most likely operate in teams of professionals of various and diverse professional disciplines. Therefore, higher education programs increasingly focus on students developing those skills. This paradigm shift in education can be described as the shift towards Smart Education, defined as education in a smart environment supported by smart technologies, making use of smart tools and smart devices [7].

All the innovations in education, such as massive open online courses or MOOCs, ICT-in-education, blended learning, flip-the-classroom and student-driven learning, can make the didactics in classrooms more attractive, but they come with a price for the support service. The increasingly diverse range of didactical concepts that are being applied in the classroom, put a greater burden on the organization in both support and management services. More flexibility as a result of different didactics make it harder to anticipate and plan. As the new or smart education results in a much more diverse variety of lectures, it also brings uncertainty to lecturers and program managers, who hedge this uncertainty by overclaiming teaching facilities, to ensure coverage in all cases. In many educational organizations, management and support work is still being done according to the principles of the "first industrial revolution" [6]. The organization works with batches of students, is cohort focused, with the same start- and end times for all student groups, with classes at fixed locations, with the timetable already fixed far in advance, and with strictly regulated consultations for all educational methods - be it lectures, seminars, practicals or projects. Support services responsible for facilities or educational timetabling, try to match education activities with appropriate facilities. In this context, the word "appropriate" should mean that the best facility should be found to fit the didactical form of the lecture. However, in our experience, timetabling rarely looks beyond the required number of seats, and handles at most a handful of lecture types. The increasing diversity of lectures brought along by education 2.0 necessitates a closer look at the fit between facilities and a lecture type or didactical form, and thus poses a greater burden on the organization services. Neglecting this shows the lack of adaptiveness in the education organizations, which can drastically hamper educational innovations.

This all supports the call for coordination by planning and control of the education organization. Changes towards smart education and the consequential increased complexity require adaptation of the education organization such as administration, HR, facility- and ICT-management. An internal report made by a consultancy company in The Netherlands supports this conclusion. For several performance factors they researched at various education organizations, the analysis not only showed gaps between strategy and operation, but, more pressing, gaps in coordination between different processes or activities are serious [8]. The planning of tasks on strategic, tactical and operational level does exist within each process, however, the coordination of the planning between those processes is insufficient. E.g. the tactical planning in the support process does not necessarily fit with the tactical planning in the primary process. This means that the level of adaptation of the organization is not smart enough for an optimal alignment between process areas, leading to unnecessary obstructions of innovation.

2 Planning and Control Models: Literature Review

This section examines various sectors to find frameworks for coordinating processes in organizations. Several P&C models found in theory are described, ranging from general to specific applied models in different fields. The main goal is not to discuss the models in length, but to understand the application of the models and to translate it to an educational setting.

Definition of Planning and Control. The purpose of planning and control is '*to ensure that the operation runs effectively and produces products and services as it should do*' [9], the realization of a certain delivery at minimal costs, or simply the fitting of supply and demand. Planning is the intellectual anticipation of possible future situations, the selection of desirable situations to be achieved by setting objectives, and the determination of relevant actions that need to be taken in order to reach those objectives at a reasonable cost. In other words, planning implies thinking about the future and trying to assume control over future events by organizing and managing resources so that they cater to the successful completion of the objectives set forth [10]. Educational planning, in its broadest sense, is the application of rational, systematic analysis of the process of educational development with the aim of making education more effective and efficient in responding to the needs and goals of its students and society [10]. Educational planning and control can be compared to P&C systems used for other organizations, such as production and service organizations. Research on P&C models shows a large number of these systems, many of which are grounded on the same basic principles of a process- or information model, translating strategic policies to operational activities through tactical management and coordination.

General Planning and Control Models. A P&C model is designed to efficiently manage the flow of materials, the utilization of people and equipment, and to respond to customer requirements by utilizing the capacity of suppliers, internal facilities and in some cases of customers to meet customer demand [11]. One of the first hierarchical planning and control models was proposed by Anthony [12], as shown in Fig. 1. This early model of P&C consists of three hierarchical levels, from strategic through tactical to the operational level. The strategic level is concerned with aggregate decisions, e.g. the total capacity required. These decisions are disaggregated in a later stage, e.g. into specific capacity allocation plans on operational level. Strategic plans are often called long-term plans, tactical plans medium-term and operational plans are called short-term plans. There is no standard length for each term.

Fig. 1. Anthony's framework for management control (1965)

Later, in 1974, De Leeuw [13] presented his governance model, as given in Fig. 2. This model emphasizes the relations between environment, governing body, the controlled system and system boundaries. Control of the organization is a result of coordination between both the governing body as well as the controlled system.

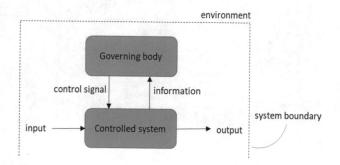

Fig. 2. Model of De Leeuw (1974)

In other models, distinctions are made between the different areas of expertise to produce the goods or services. The Porter value chain [14] was introduced in 1985, and shows the strategically relevant activities to gain insight into the possibilities of operating better or more efficiently. The value chain shows that we should not approach the company as a whole to achieve competitive advantage, but that we must have insight in our primary and supportive activities.

More recently, in 2010, Slack et al. [15] presented an extended model for P&C in production organizations, as shown in Fig. 3. This model elaborates on the previous models described, by incorporating the input and output streams in the organization.

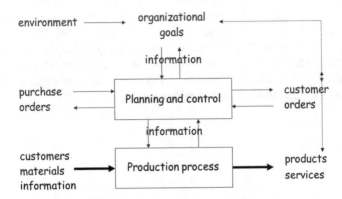

Fig. 3. Model Slack et al. for supply chain management (2010)

ISA-95, displayed in Fig. 4, is an international standard from the International Society of Automation [16] for developing an automated interface between enterprise and control systems. The objectives of this system are to provide consistent terminology as a foundation for supplier and manufacturer communications, provide consistent information models, and to provide consistent operations models as a foundation for clarifying application functionality and how information is to be used.

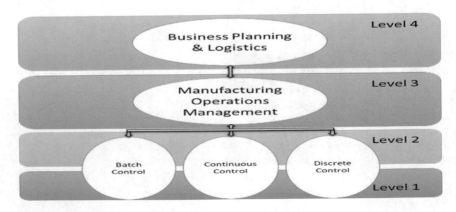

Fig. 4. ISA95 model for connecting enterprise and control systems

A widely accepted premise is that every organization has three main types of processes: primary processes addressing the customer, support processes to facilitate the primary process, and directive processes to manage both other processes, as layers in an hierarchical decomposition. The basic layout is a generic template within which each organization has its own variations. Primary processes are considered the most distinctive. Figure 5 shows these processes in relation to each other [17].

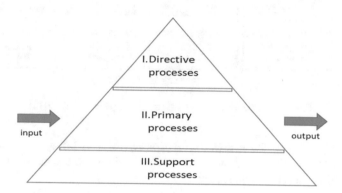

Fig. 5. Three main types of organization processes (2003)

Our last generic model for organizational coordination described here, is the classic division of the organization by structure, as given in Fig. 6. It is the well-known model for task allocation, coordination and supervision [18].

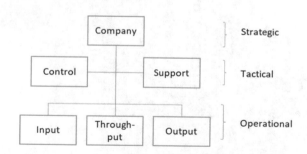

Fig. 6. Hierarchical organization structure (1941)

Applied P&C Models. In healthcare, the framework for healthcare planning and control, as proposed by Hans, Van Houdenhoven, and Hulshof, is given in Fig. 7 [19]. It splits the P&C process into different managerial areas and hierarchical levels and describes the contents of the processes in each area on the different levels.

Models for education primarily concern the education itself, not the organization of it. However, one model was developed for P&C in education by the SURF special interest group on Education Logistics in the Netherlands, as shown in Fig. 8 [3]. The main target of this model is to give a better understanding of all aspects of logistics in

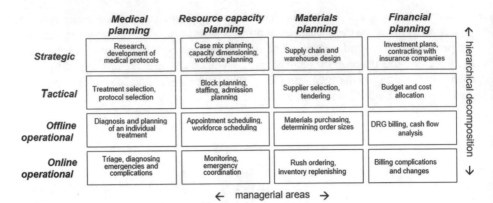

Fig. 7. Framework for healthcare P&C (Hans et al. 2012)

education. By bringing those aspects in perspective as a honeycomb pattern, it is supposed to work as a puzzle. It does not automatically fit, so discussion is encouraged between educational and support staff.

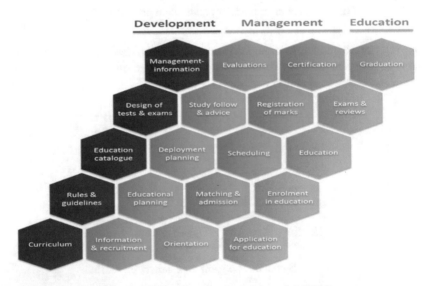

Fig. 8. SURF education logistics model (2014)

Usability of the Models. Each model described has elements that can be valuable for P&C in education to make the organization smart and adaptable. The search for more models does not necessarily lead to better or new insights, but to more of the same.

Many general P&C models are aimed at production. However, education differs from production on essential aspects. Production is concerned with producing goods or services, whereas education is aimed at facilitating the learning processes of students. Its input is people with their own opinions, wishes and learning curves, while the input for production is means. The throughput of education is the transfer of knowledge and competences through theory and practice assignments, not handling materials as in production. The customer wants to learn and develop. And last, the output of education is added value in the form of a higher level of knowledge and skills, instead of added value in the form of an end product or service. The customer itself is therefore both the product that goes through the production process of education, and the customer of the process.

The P&C models of Anthony and De Leeuw give a good starting point as a governance model for education logistics, as they address the need for coordination on and between different hierarchical levels, but they are not specific enough to coordinate the education organization. They also do not zoom in on coordination between different process areas.

Both the model of De Leeuw and the model of Slack et al. describe the production process from an 'helicopter view', with the production process as a black box. However, to translate the model to a model for the coordination of the education organization, being able to look inside the production processes of education is required.

The ISA95 standard is a good guide to mapping which information needs to be exchanged between office automation systems and production automation systems [20]. However, the ISA95 model does not describe hierarchy or timeframes. It describes the flows of information. ISA95 can be applied in educational setting to exchange information, but only if the different systems that the model uses can be distinguished in education. They should be specified in order to be useful for a P&C model in education.

The hierarchical organization structure model does not clarify P&C in the organization, but does emphasize the importance of hierarchy in organizational coordination.

The healthcare P&C model from Hans et al. [19] seems suitable as a model for educational coordination, because of the distinction between different management areas and division of hierarchical focus. This model can apply because, just as in education, the customer is input, throughput and output of the primary process. The misfit of this model as a model for education logistics is that education is not aimed at the care for one disease or one treatment at a time, but at a complete range or route.

The education logistics model provided by SURF [3] has no interaction with other settings as it is only aimed at the own organization. The education logistics model can be considered as not finished, as the connections between the cells are not described and the hierarchical levels are not recognizable. In the description of the model it is stated that the model gives a starting point for discussion and is not intended as a scientific model.

All of the models described are examples of P&C models, ranging from foundations of general P&C models to P&C models for specific fields. The idea of a student-as-customer in education 2.0 differs from the idea of customers in most fields. The comparison with production seems suitable at first glance, but differs at crucial points. All models discussed have one or more elements missing in order to function as a suitable framework for P&C in smart education organizations. None of them focuses on the adaptability of education organizations, as they do not enable the organization to respond to the requested changes. Therefore, we introduce a new model as P&C model for education.

3 Planning and Control in Education

A suitable framework for P&C in education organizations would use the abstraction of a general model and combine it with specific content to control the different processes in education organizations in order to support a higher level of smartness. First, a distinction is made between the primary processes, the management processes and the support processes. Second, a distinction is made between the hierarchical levels to overview and coordinate differences in time horizons. Last, the framework must broaden the discussions about P&C to transcend the connotation of finance to be able to improve on the performance of the organization.

Setting Up the P&C Model for Education. From these constraints, a governance model for education logistics can be drawn. Coordination in the educational organization is planned on three hierarchical levels. Strategic decisions in educational planning are less specific and indicate broad directions of development, for instance the decisions of increasing or decreasing total capacity or resources (e.g. workforce), offering additional education, curricula or studies. The planning horizon for these decisions is mostly related to the strategic business plan of the organization. Tactical decisions in educational planning consist of the allocation of resources available for education. Typical decisions are the amount of lecture rooms needed in each period, or the number of contact hours needed in each period. Also, the levels of workforce and number of lectures to be used are tactical decisions. To make effective tactical decisions it is generally sufficient to consider aggregate data. The basic operational decisions of educational planning consist of scheduling the lectures or contact hours in each timeframe and the corresponding lecture rooms and lecturers needed. Operational decisions are made subject to the limitations imposed by the tactical level and require a high degree of detailed information. The planning horizon is often related to the mapping of the academic year in semesters and timetable of the students.

Framework for P&C of Education Organizations. The educational P&C framework consists of three process areas: Educational planning, Support planning and Management or Directive planning. The primary process planning consists of decision making about the development of education: curricula, didactics, lectures, exams etc. Predictable activities are planned centrally and at long term, whereas unpredictable information are decentralized and short term. Support processes concern resource capacity planning, which consist of facilities, equipment and staff. Directive processes concern management, controlling, and financial planning such as cost and revenues. The framework is given in Fig. 9.

The usability of this model lies its simplicity yet broadness. The alignment between the process areas is precisely what is needed of coordination in the smart organization. Continuous alignment is needed between the process areas to match strategy, tactics and operations. A specific completion of the model for the coordination in an education organizations would look as shown in Fig. 10.

Process:	Directive	Primary	Support
Timeframe:			
Strategic planning	Positioning and capacity	Control of delivering courses and programs	Quality of support levels
Tactical planning	Budgetting and staffing	Set-up of education	Capacity of support
Operational planning	Checks and balances	Education	Administration and deployment

← — — — process areas — — — →

hierarchical decomposition ↑↓

Fig. 9. The P&C framework for the education organization

Process / Timeframe	directive processes: management	primary processes: education	secondary processes: support
Strategic planning	• Positioning of the organisation • Image building • Building plans and housing • Funding plans for research, education and support	• Education level (academic or applied sciences) • Education form (fulltime, part-time, dual) • Type of education (bachelor, master, associate degree) • Educational programmes • Number of study years • Titles of programmes • Required competences • Cycles of accreditation • Educational form	• Quality level of HR, IT and other support service • Capacity of support service • Sustainability and deployment of support • Education locations

Tactical planning	• Organisation: budgeting and staffing:
	  • Distribution of staff
	  • Amount of teaching and research plans
	• Focus points
	• Yearly budgets
		• Majors and Minors
		• Certifications of courses/programmes
		• Educational staff planning
		• Educational accreditations
		• Curriculum
		• Education plans
		• Exam plans
			• Year plans of support (IT, HR)
			• Fixed capacity plans
			• Expected variable (flexible) capacity plans
			• Certifications

Operational planning	• Administration • Budgets • Checks & balances • Training • Planning and control • Tracking and monitoring	• Courses • Exams • Allocation of staff • Allocation of rooms • Training • Quality checks	• Administration • Deployment of staff • Deployment of locations • Training • Checks on the functioning of automation, facilities, procedures

Fig. 10. Example completion of framework for education P&C

What facilities are needed in the classrooms, what size, what decoration, what locations in what building, etc., are all examples of questions that can only be answered if cooperation between the process areas is present and structured. Comparable questions can be issued by the ICT services department, and again the answer is a broader alignment.

4 Discussion and Further Research

In our own experience, people talking about the strategy of the education organization refer to the board of directors, while on the other hand, people talking about operations refer to the educational staff. Our model shows that this mind-set can and should be broadened. Each of the process areas has their own strategic, tactical and operational planning. This broadens the discussions to enable the organization to adapt to the new demands from smart education.

As we have shown, none of the existing models in manufacturing, production, government, health, education or any other areas, fit the requirements that we pose upon a suitable model for P&C in education. The new framework for education organizations to govern their logistics, fits the new paradigms emerging in smart education. It mitigates the flaws in P&C models in other fields and can therefore lead to better coordination and adaptiveness. Further research must show that education organizations following this framework, gain on efficiency and coordination in the smart organization.

References

1. Dima, A.M.: Challenges and Opportunities for Innovation in Teaching and Learning in an Interdisciplinary Environment, Knowledge Management Innovations for Interdisciplinary Education: Organizational Applications: Organizational Applications, p. 347 (2012). ISSN 1466619708
2. Tan, J.C.: Project-based Learning for Academically-able Students: Hwa Chong Institution in Singapore. Springer (2016). ISBN 9463007326 red
3. SURF: Onderwijslogistiekmodel: Beter communicern door gemeenschappelijke taal. SURF, Utrecht (2014)
4. Keats, D., Schmidt, J.P.: The genesis and emergence of Education 3.0 in higher education and its potential for Africa. First Monday **12**(3), 3 (2007)
5. Hans, E.W.: Is Better Now, Doctor? University of Twente, Enschede (2015)
6. sir Robinson, K.: Changing Education Paradigms. TED talks (2010)
7. Coccoli, M., Guercio, A., Maresca, P., Stanganelli, L.: Smarter universities: a vision for the fast changing digital era. J. Vis. Lang. Comput. **25**(6), 1003–1011 (2014)
8. Nijhuis-Boer, V.: Analysis 7S Models. Enschede (2017)
9. Slack, N.: Generic trade-offs and responses: an operations strategy analysis. Int. J. Bus. Perform. Manag. **1**(1), 13–27 (1998)
10. Unesco: Strategic Planning: Concept and rationale. International Institute for Educational Planning, Paris (2010)
11. Vollmann, T.E.: Manufacturing Planning And Control Systems For Supply Chain Management: The Definitive Guide for Professionals. McGraw-Hill Education (2005). ISBN 9780071440332
12. Anthony, R.N.: Planning and Control Systems: A Framework for Analysis, Division of Research, Graduate School of Business Administration, Harvard University (1965)
13. de Leeuw, A.C.J.: Systeemleer en organisatiekunde: een onderzoek naar mogelijke bijdragen van de systeemleer tot een integrale organisatiekunde (1974)
14. Porter, M.E.: Competitive Advantage: Creating and Sustaining Superior Performance. Free press, New York (1985)
15. Slack, N., Chambers, S., Johnston, R.: Operations Management. Pearson education (2010)
16. The International Society of Automation: ISA95, Enterprise-Control System Integration. Research Triangle Park, NC 27709, n.d.
17. Nieuwenhuis, M.A.: The Art of Management, ISBN-13: 978-90-806665-1-1 red., the-art.nl, 2003–2010

18. Davis, R.C.: The influence of the unit of supervision and the span of executive control on the economy of line organization structure. Bureau of business research, The college of Commere and administration (1941)
19. Hans, E.W., Van Houdenhoven, M., Hulshof, P.J.H.: A Framework for Healthcare Planning and Control. In: Handbook of healthcare system scheduling (2012)
20. Wikipedia: ANSI/ISA-95, 17 November 2015. https://nl.wikipedia.org/wiki/ANSI/ISA-95

Sustainable Learning Technologies:
Smart Higher Education Futures

Proposing an Innovative Design Based Evaluation Model for Smart Sustainable Learning Technologies

Madhumita Bhattacharya[1,2](✉) and Steven Coombs[3]

[1] New Paradigm Solutions Ltd., Palmerston North 4410, New Zealand
mmitab@gmail.com
[2] Global Association for Research and Development of Educational Technology,
Victoria, BC, Canada
[3] Hamdan Bin Mohammed Smart University, Dubai, UAE
s.coombs2015@gmail.com

Abstract. The proposed evaluation model has been created from the principles of adopting a Design Based Research (DBR) approach. The DBR process has been reviewed in terms of its current limitations and then extended into a new paradigm and academic epistemology. The proposed framework explains the nature of all design-based phenomena and provides an epistemological postulate for a Design Based Evaluation research design and methodology. In creating this model the authors have considered various theories and research methods and methodologies e.g. Activity Theory, Actor Network Theory, Grounded Theory, Collaborative Protocol Analysis, Evaluation of Self-reflection. This synthesis of 'theories' has provided a structure in the otherwise unelaborated empty shell of DBR. Using the proposed Design Based Evaluation (DBE) Model it is possible to introduce and evaluate new learning technology innovations and at the same time explore its potential and make improvements for long-term implementation and sustainability. The authors in this article suggest that continuous improvement of technologies and support systems are needed in order to move forwards and not to lag behind, thereby ensuring a future of smart sustainable learning technologies.

Keywords: Design based research · Design Based Model
Design based phenomena · Sustainable learning technologies · Activity theory
Actor network theory · Ethnographic research · Grounded theory
Collaborative protocol analysis · Evaluation of self reflection
Evidence-based authentic learning

1 Background of Design Based Research (DBR)

1.1 DBR's Conceptual Evolution

In this section the authors introduce and review the concept of design-based research (DBR) [1] with key ideas and a new definition of what it represents to the field of research methodology and why this new paradigm is important for sustainability and long-term social impact [2]. The authors also maintain that DBR can be used to analyze any research problems not just those suited for evaluation of technologies, i.e. that DBR represents its own type of research methodology. However, the DBR approach through participant engagement and evaluation can be used to oversee the introduction of new technologies in general and in particular the authors consider a focus on learning technologies linked to managing the change process on a sustainable basis for leveraging useful social impact. Currently, the problem (gap) is that many technologies are introduced with little thought of the wider social impact, or with any testing/ evaluation by the primary users, e.g. students, patients, customers etc. This DBR prerequisite to include primary users as research participants is an argument that extends to developing a research policy for identifying a common set of ethics and shared values for users of the technology innovations introduced.

The DBR approach has been proposed since the early 1990s [3, 4], but presents challenges and limitations which need to be better understood if it is to meet its long-term potential as a stand-alone research methodology [5, 6]. DBR has not met its full potential due to many limitations, but the authors maintain it is due to the lack of a proper DBR research methodology based upon an epistemology with an underpinning academic framework. This represents a significant gap in the field of research methodology given its potential for meaningful social impact research. This goal has been elusive due to the multi-disciplinary nature of the otherwise tentative and shaky conceptual framework that DBR has so far been defined. Limitations include definitions regarding how to organize the research setting, roles of participants, number of iterative cycles linked to the methodology of the reflective evidence etc. Thus, a system-based version of DBR is required and the way that reflection in real-life settings can be made more systematic and properly recorded as part of the DBR qualitative research process. This qualitative approach also needs to be made objective so as to also allow the analysis of quantitative data suggesting DBR settings require a new type of mixed methods as part of the research strategy. Activity theory [7] and actor network theory [8] consider that human beings and their socio-physical interactions with physical objects operate as essential and symbiotic parts of a common research design. The question is what are the assumptions underpinning the concept of "Design" in DBR? What is Design-based relative to within the research paradigm of DBR? What is the nature of *design* within a research setting? Thus, if the design is relative to learning then DBR needs to be relative to learning design [9].

Likewise, design might be relative to the living persons within a research setting including cultural and ethnographic values [10], suggesting that the organizational nature and value-systems of the research participants is also part of the design rationale [11]. This suggests a design-based research paradigm that includes as one of the key components cultural and ethical values that extends the horizons of what is meant by a design-based approach.

The authors propose the nature of the design paradigm for DBR relative to the phenomena being investigated with the following postulate:

The research paradigm for design-based research requires an identification and systematic mapping of a common set of overlapping domains that includes personal and group participant learning, participant cultural and ethical values, personal commitment of time and availability of other required resources (human and physical as prerequisites), and group agreement on the potential social impacts and recognition of contributions for lifelong learning sustainability.

The by-product of DBR is human understanding and knowledge sharing as part of the living research process and is both a major benefit and purpose of leveraging such a social impact. At the same time it is intended participants engaged with a DBR project will be improving a wide skill set and personal attributes including, but not limited to: collaborative social and learning skills; communication skills; interpersonal skills; cultural empathy and responsiveness; self-organizing capability linked to personal management; adaptability to change; and, gaining confidence with new technologies [12].

Through the inclusion of key participants affected by the change process of the introduction of new systems and technology represents a major strength of the DBR paradigm and the elaboration of their involvement as part of the validated design of the research experiment. This strength is also achieved through adopting a systematic DBR evaluation approach that enables participants to both learn and adapt from change. An additional strength is that the DBR process includes the aspect of reflection by all the participants throughout the various iterative cycles, operating as a key characteristic of the whole project.

1.2   DBR as an Evaluation Process

So far the authors have reviewed the conceptual nature of DBR introducing the idea that DBR can potentially operate as an evaluation model for many practical systems including emerging and extant learning technologies. The authors postulated [2] that DBR represented a processed-based model to support sustainable learning technologies:

The design based research process can act as a filter to all new and innovative technologies for determining the future sustainability, improvement and ultimate useful application of emerging technologies (especially learning technologies).

The authors propose that DBR offers a suitable evaluation process for applied research that we will now refer to as design-based evaluation (DBE). Design-based evaluation represents a framework with a <u>unique set of characteristics</u> and features that focus on 4 main components for any particular sustainable technology under review:

1. Formation of a collaborative team to manage and test any proposed change. Thus, field-based evaluation with users as participants to the process can be linked to practitioner research e.g. for professions including medical, banking, business and education etc.
2. The characteristics of DBR evaluation also includes providing a rationale for analyzing the need for change in the first place; why are we seeking new solutions and for what defined improvements and benefits? What about the extant technology and its purposes? In response an analysis of extant technologies is required to articulate the feasibility of the proposed change and represents a long or short-term solution and identifies the justification for any short term fix?
3. The third characteristic is the nature of DBE as an iterative process of testing the application of any new technology in practice and determining its overall sustainability.

4. The nature and role of systematic critical reflection to fully engage all participants, so as to provide their feedback and suggested improvements before, during and after the DBE process. The DBE process is one where the continuous reflection during the various iterative DBE cycles provides the resultant review evidence and analysis of the effectiveness of the scrutinized technology for potential future sustainable implementation.

The potential applications for the above new paradigm of DBE is wide and far. One application of DBE might include areas of work-based practitioner research as teams focusing on the investigation of available new technologies for suitability, sustainability and affordability. Another application could be an assessment of new technology by members of the public for reviewing its suitability prior to being fully implemented to support and sustain their own community e.g. technology to sustain the learning of endangered indigenous languages and cultures. Thus, DBR becomes transformed into a general Design Based Evaluation (DBE) process that can be modeled to manage technological and other change including the primary users and instigators of any community or business project e.g. an international development project to support literacy in remote island communities.

Another avenue for DBE applications is in the implementation of any new technology supported by critical thinking and reflective e-tools. For example, Bhattacharya and Hartnett [13] stated that the concept of digital or eportfolios goes beyond the archiving of simple text and still images. Instead, one can incorporate multimedia to demonstrate new knowledge and skills acquired by the user as learner. Therefore the digital realm of the eportfolio is being extended as new tools and technologies are developed and included to support *evidence-based authentic learning* anchored and personalized through deeper reflection and richer engagement. This suggests that eportfolio development is not only about a "collection" of artefacts, but instead can help to articulate evidence of learning and "reflection" on the processes being engaged within. The eportfolio therefore produces more than just output of learning and operates as a flexible transformative learning technology, where the primary user is engaged with meaningful "interactions" as critical reflection with consequent transfer of learning suitable for lifelong learning. In this "networking age" no learning can be labelled as merely independent and individual. Knowledge is capable of being negotiated and distributed among people and artefacts using a range of potential learning technologies that represents a sustainability of such learning environments [14]. This also supports the concepts of activity theory and actor network theory.

What the authors have noticed is that a design-based approach can be applied to many different situations and social scenarios of applied research that requires systematic thinking, modelling and evaluation. This implies an application towards most phenomenological situations with real life participants where a design-based approach would help to *systematically organise* and conceptually *model* most forms of social research. This suggests a universal case for understanding the nature of any design-based phenomena operating within a DBR paradigm.

Consequently, the authors wish to propose the following postulate for design-based phenomena:

That all design-based phenomena are a systems thinking blueprint representing a conceptual framework of all the identified and relevant processes and components involved.

2   Introduction of a Design Based Evaluation (DBE) Model

2.1   Introducing the Framework for DBE

As discussed previously the authors provided a rationale for choosing a DBR framework and being engaged in a deeper scrutiny to develop the DBE process. The authors have justified in their research [2] how DBE could be used to benchmark the sustainability of any technology. The authors are now suggesting how the new DBE model integrates the DBR methodology along with the elements and processes of Activity Theory, Actor Network Theory, Ethnographic Research Methodology, Grounded Theory, Collaborative Protocol Analysis, Rubric for Evaluation of Reflection and guidelines for the learning analysis of blogs.

The following items integrate in Gestaltian fashion the epistemological components of the proposed design-based theoretical framework:

Activity theory: is all about 'who is doing what, why and how'. In Activity Theory, the relationship between subject (human doer) and object (the thing being done) forms the core of an activity. The object of an activity encompasses the activity's focus and purpose while the subject, a person or group engaged in the activity, incorporates the subject's/s/s' various motives. The outcomes of an activity can be the intended ones, but there can also be others that are unintended. Sometimes referred to as the Cultural-Historical Activity Theory (CHAT), [15].

1. Actor network theory: ANT is a constructivist approach in that it avoids essentialist explanations of events or innovations (i.e. ANT explains a successful theory by understanding the combinations and interactions of elements that make it successful, rather than saying it is true and the others are false). Likewise, it is not a cohesive theory in itself [16]. Therefore any technology could be sustainable if tested through iterative cycles of DBR by various groups in a number of different setups.

2. Ethnographic research and Grounded Theory [17]. The emphasis in ethnography is on studying an entire culture. Originally, the idea of a culture was tied to the notion of ethnicity and geographic location (e.g., the culture of the Trobriand Islands), but it has been broadened to include virtually any group or organization. That is, we can study the "culture" of a business or defined group (e.g., a Rotary club). The most common ethnographic approach is participant observation as a part of field research. The ethnographer becomes immersed in the culture as an active participant and records extensive field notes. As in grounded theory, there is no preset limiting of what will be observed and no real ending point in an ethnographic study. Grounded theory is a complex iterative process [18].

3. The approach of collaborative protocol analysis [19] provides further insight into the evaluation of the technology from the usability point of view as well as learning from each other.
4. Participant's engagement in the DBE process could be using social networking tools such as blogs. Therefore one has to evaluate the blogs administering the coding mechanism advocated by Grounded Theory and developed by Bhattacharya [20].
5. Rubric for Evaluation of Reflection [21, 22] at the different iterative cycle of the DBE.

Synthesis and drawing out the relevant procedures from all the above methodologies embedded in the theories manifests with the insight of the DBR processes. This then underpins the evolution of the DBE, which is a new framework of thinking operating as a conceptual archetype for Design Based Modelling (see Fig. 1).

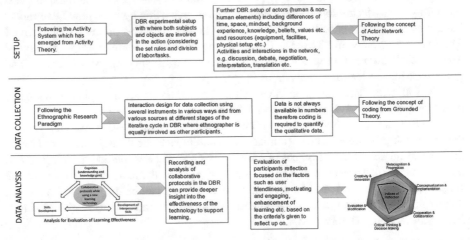

Fig. 1. Procedures and epistemologies of a design-based evaluation framework

The Fig. 1 DBR and DBE framework is subject to an iterative cycle [23] based on repeating the 3 phases of setup, data collection and analysis with a subsequent review and modification prior to the start of each cycle.

2.2   DBE for Learning Technology

The authors have investigated several existing technology evaluation models for the purpose of benchmarking them against the proposed DBE framework relative to their suitability to analyse and evaluate sustainable technologies using a DBR iterative cycle. The authors could only identify two models that were partially relevant for this purpose: (1) SAMR 4-step model for improving learning through using technology; and, (2) Owston's 2007 proposed sustainability of innovative use of technology in the classroom.

The most widely used model is called SAMR (Substitution Augmentation Modification Redefinition) (Fig. 2), which is mainly used for evaluating technology for its use in improving learning, teaching and assessment, e.g., Work [24] provided details of applying the SAMR model to enhance learning using readily available tools and technology in the classroom and beyond. Starting with a PowerPoint slideshow in the classroom (Substitution), to web-based Prezi presentation (Augmentation), to enhanced previous presentations with video, audio, hyperlinks, etc. (Modification) and finally using the Nearpod (an interactive presentation and assessment tool for flexible use with mobile apps.) presentation (Redefinition). In another study Romrell et al. [25] reported how mobile devices can be used to improve learning using the SAMR model. They provided examples of learning activities that fall within each of the four classifications of the SAMR Model. The SAMR model provides a pathway to enhance and expand the features and widen the accessibility and repertoire of the available learning technologies. In order to improve this model it needs to be adapted to include the DBR characteristic of possessing an iterative cycle for ongoing evolving improvements as technology advances.

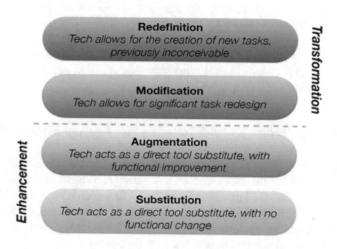

Fig. 2. The SAMR model (sourced from Schrock [26])

Another evaluation model the authors have considered is proposed by Owston [26] studied factors that contribute to the sustainability of innovative classroom use of technology. Using the Atlas.ti© qualitative software tool he mapped the relationships among codes and developed a model (see Fig. 3) that helps explain why teachers are likely to sustain innovative pedagogical practices using technology.

The Owston model in Fig. 3 has positives in that it identifies the essential and contributing factors that influence the sustainability of the innovative use of technology in the classroom. However, the downside is that this model is only applicable to physical learning environments, whereas innovative technology use has now expanded to virtual and eLearning environments. The model lacks interactive relationships and potential teamwork across the various social elements as contributing and essential

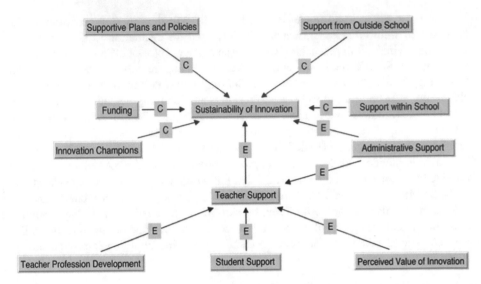

Fig. 3. Essential (E) and contributing (C) factors to the sustainability of innovative use of technology in the classroom (Sourced from Owston [27])

factors. Collaborative teamwork learning, decision making, responsibility sharing are assumptions of a DBR framework. It also lacks any iterative improvement cycle as in the case of the previous model reviewed.

3 Concluding Remarks and Insights for Future Research

The strength of design-based research is its flexibility and applicability towards any phenomena and can be the value-add of any resultant synthesized research methodology leading to an enriched research paradigm. This research paper has identified and linked DBR with DBE and DBM, where design-based research represents the research processes, design-based evaluation the research methodology and design, and the design-based model as the conceptual research framework.

The design-based approach provides a continuous process for longitudinal research that implies a sustainability agenda for enabling naturalistic and authentic living research paradigms. The DBR and DBE research methodology and approach sees on-going improvement as part of its natural evolution and sustainable change agenda for action. This requires research into new types of long-term qualitative and quantitative data that fields such as learning analytics can bring, whilst also ensuring the long-term robustness of this research paradigm for generalizability, social transfer and wider impact.

At the same time the design-based modeling approach associated with DBR can be applied to learning environments linked to sustainable change. A design-based evaluation model can therefore be used to both improve and continuously develop learning technologies for sustainability in whatever format or environment that they might appear and forms an essential future research agenda.

References

1. Amiel, T., Reeves, T.C.: design-based research and educational technology: rethinking technology and the research agenda. Educ. Technol. Soc. **11**(4), 29–40 (2008)
2. Bhattacharya, M., Coombs, S.: A design-based research approach for educational innovation: sustainable learning technologies as the alternative holistic paradigm antithesis of disruptive technologies. In: Proceedings of Innovation Arabia 11, Dubai, UAE, 11–13 March 2018
3. Collins, A.: Toward a Design Science of Education. Center for Technology in Education, New York (1990). http://cct2.edc.org/ccthome/reports/tr1.html. Accessed 15 Mar 2018
4. Collins, A.: Toward a design science of education. In: Scanlon, E., O'Shea, T. (eds.) New Directions in Educational Technology, pp. 15–22. Springer, New York (1992)
5. Kelly, A.E.: Design research in education: yes, but is it methodological? J. Learn. Sci. **13**(1), 115–128 (2004)
6. Anderson, T., Shattuck, J.: Design-based research: a decade of progress in education research? Educ. Res. **41**(1), 16–25 (2012). https://doi.org/10.3102/0013189X11428813
7. Nardi, B. (ed.): Context and Consciousness: Activity Theory and Human-Computer Interaction. MIT Press, Cambridge (1996). ISBN 0-262-14058-6
8. Latour, B.: On actor-network theory: a few clarifications, Soziale Welt, vol. 47. Jahrg., H. 4, pp. 369–381 (1996)
9. Bannan-Ritland, B.: The role of design in research: the integrative learning design framework. Educ. Res. **32**(1), 21–24 (2003)
10. Coombs, S., Potts, M., Whitehead, J.: International Educational Development and Learning through Sustainable Partnerships: Living Global Citizenship. Palgrave Macmillan, London (2014)
11. Oubenaïssa-Giardina, L., Bhattacharya, M.: Managing technological constraints and educational aspiration in a multicultural E-learning environment design. J. Interact. Learn. Res. **18**(1), 135–144 (2007). Association for the Advancement of Computing in Education (AACE), Waynesville. https://www.learntechlib.org/p/21932/. Accessed 15 Mar 2018
12. Bhattacharya, M., Mach, N., Moallem, M.: Editing an electronic book in virtual team: our journey. In: Ho, C., Lin, M. (eds.) Proceedings of E-learn 2011–World Conference on E-learning in Corporate, Government, Healthcare, and Higher Education. Association for the Advancement of Computing in Education (AACE), Honolulu, Hawaii, USA, p. 2569 (2011). https://www.learntechlib.org/p/39115/. Accessed 15 Mar 2018
13. Bhattacharya, M., Hartnett, M.: E-portfolio assessment in higher education. In: Proceedings of the Frontiers in Education Conference, Milwaukee, pp T1G-19–T1G-24 (2007)
14. Bhattacharya, M.: Sustainability of learning in 21st century: handle the present and lead the future. In: Bonk, C., Lee, M., Reynolds, T. (eds.) Proceedings of E-learn 2008–World Conference on E-learning in Corporate, Government, Healthcare, and Higher Education, pp. 260–266. Association for the Advancement of Computing in Education (AACE), Las Vegas (2008). https://www.learntechlib.org/p/29615/. Accessed 15 Mar 2018
15. Hasan, H., Kazlauskas, A.: Activity theory: who is doing what, why and how. In: Hasan, H. (ed.) Being Practical with Theory: A Window into Business Research, pp. 9–14. THEORI, Wollongong (2014). http://eurekaconnection.files.wordpress.com/2014/02/p-09-14-activity-theory-theori-ebook-2014.pdf. Accessed 15 Mar 2018
16. Simandan, D.: Competition, contingency, and destabilization in urban assemblages and actor-networks. Urban Geogr. (2017). https://doi.org/10.1080/02723638.2017.1382307
17. Glaser, B., Strauss, A.: The Discovery of Grounded Theory: Strategies for Qualitative Research. Transaction Publishers, New Brunswick (1967)

18. Trochim, W.M.: The Research Methods Knowledge Base, 2nd edn. (2006). http://www.socialresearchmethods.net/kb. Accessed 15 Mar 2018
19. Bhattacharya, M., Akahori, K.: Effectiveness study on Design and Development of Visual (DDV) using protocol analysis. In: Joint Conference on Educational Technology, Tokyo, pp. 407–408 (1997)
20. Bhattacharya, M.: Tools for Analysis of Research Data Collected from Wikis and Blogs, Canadian Institute for Distance Education Research (2008). https://auspace.athabascau.ca/handle/2149/1484/browse?type=author&value=Bhattacharya%2C+Madhumita.   Accessed 15 Mar 2018
21. Bhattacharya, M.: Electronic portfolios, students reflective practices, and the evaluation of effective learning. In: Australian Association for Research in Education Annual Conference, AARE 2001, Fremantle, Australia (2001). https://www.aare.edu.au/data/publications/2001/bha01333.pdf. Accessed 15 Mar 2018
22. Bhattacharya, M., Novak, S.: A integrative model for the evaluation of E-portfolio. In: 7th IEEE International Conference on Advanced Learning Technologies, Niigata, Japan, pp. 215–216 (2007). https://pdfs.semanticscholar.org/ef7f/9599e55b2b9200383e7975a603844f2df33d.pdf. Accessed 15 Mar 2018
23. Bhattacharya, M., Dron, J.: Mining collective intelligence for creativity and innovation: a research proposal. In: Siemens, G., Fulford, C. (eds.) Proceedings of ED-MEDIA 2009–World Conference on Educational Multimedia, Hypermedia & Telecommunications, pp. 3514–3519. Association for the Advancement of Computing in Education (AACE), Honolulu (2009). https://www.learntechlib.org/p/31986/. Accessed 15 Mar 2018
24. Work, J.: The SAMR Model: How to Evaluate Classrooms More Effectively Technology in School, 8 August 2014. https://www.edsurge.com/news/2014-08-08-the-samr-model-how-to-evaluate-classrooms-more-effectively. Accessed 15 Mar 2018
25. Romrell, D., Kidder, L., Wood, E.: The SAMR model as a framework for evaluating mLearning. J. Asynchronous Learn. Netw. **18**(2) (2014). https://olj.onlinelearningconsortium.org/index.php/olj/article/view/435
26. Schrock, K.: SMAR and Bloom's, Kathy Schrock's Guide to Everything (2013). http://www.schrockguide.net/samr.html. Accessed 15 Mar 2018
27. Owston, R.D.: Contextual factors that sustain innovative pedagogical practice using technology: an international study. J. Educ. Change **8**(1), 61–77 (2007)

Knowledge Control in Smart Training on the Example of LMS MOODLE

Leonid L. Khoroshko[✉], Maxim A. Vikulin,
and Vladimir M. Kvashnin

Moscow Aviation Institute (National Research University), Moscow, Russia
khoroshko@mati.ru

Abstract. Though lecture materials are important, electronic smart training requires the comprehensive control of students' knowledge. Lecture materials are used for knowledge assessment in progress, which is better applied when revising the material rather than for the common knowledge control.

Keyword: Smart training · Knowledge control · e-Learning
MOODLE learning management system

1 Introduction

Testing as academic performance rating is an age-old practice. As far back as 1899, Alfredo Binetti developed a system that was indicative of problems available in children with learning and information perceiving. From then onwards, testing underwent a lot of changes. Testing has become an integral part of the teaching and learning process not only as an instrument of academic performance rating but also as a method of smart teaching in the form of training checks without limits to the number of efforts and prompts. In this process, the use of e-learning system, like LMS MOODLE, is part of a strategy for implementing smart learning tools and techniques, including e-testing.

The comprehensive knowledge control requires the division into the assessment part and the lecture part. Besides, the knowledge control shall cover not a single lecture but, for example, the full section of the course. Therefore, an individual course element to be assessed separately and provided with its own configuration is required.

The most convenient method of knowledge assessment in electronic smart training is testing. This method allows for both remote and face-to-face knowledge control. After making a single setting, the assessment will be carried out automatically without additional teacher's workload [1]. By way of example, an effort was made to develop a test system for academic performance rating in the discipline of "Data bases".

2 Question Pool Structure

For test creation, it is first necessary to form the structure of the question pool using categories of questions. Based on this, future tests will be quickly created and test questions will be continuously improved without changes in the testing structure.

© Springer International Publishing AG, part of Springer Nature 2019
V. L. Uskov et al. (Eds.): KES SEEL-18 2018, SIST 99, pp. 259–266, 2019.
https://doi.org/10.1007/978-3-319-92363-5_24

To go to the question pool, select the course, "Settings" block, "Course Management" group and click "Question Pool".

After switching to the question pool, the pool contents will display on the screen. If no questions were created in the course yet, only the drop-down list is displayed for the selection of categories and control flags for question display in the pool – it is possible to display questions from lower categories (subcategories), previous (deleted) questions and the text of the question in the list of questions (the full text of all questions is displayed).

To add the new category of questions, enter its name to be displayed in the question pool and the information on the category (optional field used for the detailed description of the subject and questions to be stored in this category). After entering all the fields, click "Add the Category", and after adding, the category will be displayed at the top of the screen in the list of categories.

Standard icons for editing the list of categories are displayed against each of the categories: "cross" to delete a category, "gear" to switch to category editing, left and right arrows to quickly manage the contents of categories of questions (without editing) [2].

3   Formation of Questions

After formation of the structure of the course subjects to be tested, proceed to the formation of questions. The simplest and most common type of questions in the system is "Multiple Choice". This type will be described below. To add a new question, click "Create a New Question" in the question pool. The screen displays the selection of the type of question (Fig. 1).

Currently, there are 19 types of test questions in the system, and this list can be added or reduced, depending on the requests from teachers. To create a "Multiple Choice" question, select this type and click "Add". The system will display the window for adding a new "Multiple Choice" question.

Fill in successively the following fields in the "General" tab:

- Question name – the brief description of the question to be displayed in the question pool;
- Default score – the score to be credited in case of the correct answer. It is recommended to use 1 exclusively because of possible difficulties in explaining the student why a different number of scores is given to one or another question. In rare cases, questions can be grouped by complexity into simple (1 score), medium-complex (2 scores) and complex (3 scores). It should be noted that the system enables the calculation of the question complexity for students based on testing results (calculated values can also be used). However, as already stated, in order to avoid conflicts questions of the same category (the same test) should have the same score;
- General feedback on the question – to be displayed to the student after his/her attempt to answer the question. Despite a certain feedback depending on the type of question and the answer given by the student, the same general feedback is displayed to all students. General feedback can be used to show students the correct answer and, possibly, the link to additional information to be used for question study;

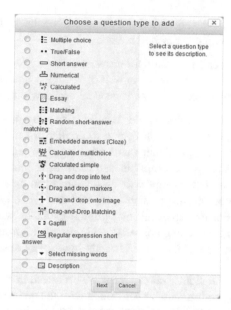

Fig. 1. Selection of the type of question

- One or more answers – setting of the number of correct answers per question (one or more);
- Random order of answers – if the parameter is enabled, the order of answers in each attempt will be random (the answers are rearranged). For certain questions containing enumerations, e.g., "variant 1", "variant 2", etc., it is recommended to disable this parameter so as not to confuse the student;
- Numbering of answer variants – by default variants are specified by letters a, b, c, d… Other variants are also possible (you can choose the preferred options).

The following categories are used for configuration of answer variants.

Text, formulas or images can be used as variants. The answer variant to be viewed by the student is entered in the "Answer" field. Grading is selected for each answer variant. If the question contains a single correct answer, one of the variants shall have "100%", if the question contains several answers, the percentage shall be divided between them. For example, the question contains 3 of 5 correct answers, then enter "33.33333%" in the "Grade" field of each of the correct variants. It is recommended to enter "–33.33333%" in the fields of incorrect variants to balance the question when using a number of test methods. The similar procedure shall apply in case of two correct answers: "50%" and "–50%".

The feedback is entered in the last field to be viewed by the student who chooses one or another answer variant. In fact, this is a feedback field. If the training testing approach is planned for application, fill in these fields as well, however, this is very labor-intensive. Therefore, control tests may have these fields empty.

If 5 variants are insufficient, you can always add them by clicking "Add 3 Answer Variants". Empty answer variants will not be displayed to students.

The next setting of the test question is a combined feedback. This setting is used for training testing. The comment can be entered:

- for any correct answer (if the student answers correctly);
- for a partially correct answer (e.g., only 2 of 3 correct answers to the question);
- for any wrong answer (e.g., 2 correct answers and 1 wrong answer).

For control test questions, these fields can be left empty, just like the penalty field for each wrong attempt. The last setting is also applied for training testing if the test enables answer correction. The penalty will be charged for each wrong answer.

The last group of settings "Prompt" is also used for training testing. In case of the first wrong attempt, the first prompt will be displayed, in case of the second wrong attempt, the second prompt will be displayed, and so on. The student's task may be gradually simplified. For example, the first prompt excludes wrong answers (i.e., the student understands his/her wrong answers), and the second prompt shows the number of correct answers (i.e., the student understands how many answer variants are to be chosen).

After entering all the fields, click "Save". The question will appear in the pool (Fig. 2). The following actions can be performed with the question by clicking standard editing icons: "gear" to switch to question settings (as specified above), "lens" to view the question (how it is seen by the student), "arrows" to move the question, "cross" to delete the question.

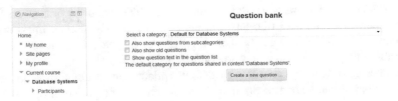

Fig. 2. Display of question in the question pool

Each question is displayed in the table that contains the following columns: "marker" to choose several questions simultaneously and handle them using buttons (delete or move to the other category), "type" in the form of the icon to show the type of the question, "question" to display the question name and editing icons, "created by" to show the author of the question, "last change" to show the last editor of the question [3].

4   Formation of Tests

After creating test questions, various tests can be quickly created – training and control tests of different types (time-limited and time-unlimited).

To create the test in the course, add the new "Test" element. After adding the test, the system will prompt to set up the test. Let's form a control test with the clear structure and the random choice of questions from the question pool as an example.

This approach allows not to change the test but to handle mainly the question pool. It is convenient for the formation of training statistics.

Enter the test name in the "General" settings group to display it in course materials.

The second field ("Introduction") is important because it contains basic information on testing for the student. The introduction shall specify the number of questions, subjects and specific requirements to answers for students (e.g., all answers shall be entered in the SI system). The system will automatically add other information to the test description so it is not necessary to enter it. It is not recommended to display the introduction on the course page, otherwise, it becomes unreadable.

The next group of settings is "Synchronization". The start and end time and date of testing can be established, and the respective information will be displayed in the course calendar. It is recommended not to establish dates and time unless you plan to continuously monitor them (during the new semester you'll have to rearrange dates). The time limit can be established in the same tab. The time of the standard control testing can vary from 30 to 60 min. Tests longer than 60 min are not efficient. The time limit for test homework can be disabled (i.e., the student can do homework as long as he/she wishes).

If the student fails to send the test attempt before the established time limit, the following situation will occur. If the student is still doing the test, the countdown timer will automatically send his/her test attempt. This parameter can have different options for the student quitting the system: ignoring the attempts not sent in due time, establishment of the grace period for sending test attempts, or automatic sending of test attempts.

The "Grading" group determines the category of grading, the number of attempts (by default, 3 attempts) and the grading method (by default is recommended). Some teachers tried different methods of grading but, finally, the "Highest Grade" method has proved to be the clearest for students (the maximum grade of several attempts is recorded in the grading log).

The "Location" settings group enables ordering of questions (e.g., grouping by pages) and navigation setting (free or successive completion of the test). It is recommended to establish these settings by defaults. It should be noted that the training system records students' answers during the navigation between pages. Therefore, if several questions are posted on the same page, data can be lost in case of possible failure of the Internet connection. Therefore, it is recommended that each question is posted on a new page.

Students can handle test questions in different ways. For example, students will answer each question without grading or feedback and see the results of the entire test after completion. This is a "Deferred Feedback" mode, which is suitable for control testing.

After answering each question, students receive an immediate feedback and if their first attempt is wrong, they immediately get the right to the second attempt with a lower grading. This is an "Interactive Mode with Multiple Attempts". These are two most common modes of question handling.

It is possible to enable an adaptive mode with or without penalties. However, the arrangement of adaptive testing is rather complex and requires separate consideration.

"Each attempt is based on the previous attempt" setting. If multiple attempts are allowed and this parameter is enabled, each new attempt will contain the results of previous attempts, which allows completing the test using several attempts. This setting is disabled for control testing and enabled for test homework. If the setting is enabled, the student always receives the questions he/she received during the first attempt and can do the homework using several attempts (Fig. 3).

Fig. 3. Test settings

The next "View Settings" group allows to configure what a student sees during the attempt, immediately after the attempt (upon completion), upon subsequent logon while the test can still be passed and cannot be passed (in case of time limits). It is recommended to leave flags in the second column only for the fields "attempt", "correct answer", "scores". The student will see these data during control testing. It is not recommended to show the correct answer to the student during control testing, otherwise, students may disclose the question database.

The "Additional Restrictions of Attempts" settings group allows for flexible restriction of the access to tests for different groups and testing technologies. For example, the test can be protected by the password or the network address, which makes it visible but not available from students' home computers.

By default, the system stipulates a 1-day delay between attempts – this value was experimentally established for students to prepare for the next test attempt. It is possible to enable the "browser security" option but it is not recommended, because this option does not actually protect students from using various test completion technologies. It is recommended to use controlled testing, i.e., in the display class under the teacher's supervision, which is much more effective than any browser security technologies, especially since students can bypass them [4].

The "Final Feedback" settings group allows giving information to the student depending on the test completion percentage. The standard comment scale is recommended: 100% – Excellent – 90% – Good – 75% – Satisfactory – 50% – Unsatisfactory – 0%. Most rating systems apply this grading scale. Percentage is filled in the "grading limit" field, and comments are entered in the "Feedback" field.

After filling in all the fields, click "Save and Show" to proceed to the formation of the test (Fig. 4).

Fig. 4. Test editing window

The central part of the screen displays the data on the maximum test grading – 100% is recommended to normalize the grading-rating course system. The test contains a blank page. The following approach to test formation is recommended. The right part of the page displays the contents of the previously formed question pool of the course. By successively choosing the test categories in the question pool, the required number of random questions from this category shall be added.

The system enables the inclusion of specific questions into the test, which means that each student will receive these questions. Therefore, it is highly probable that answers will be posted in social networks. When adding random question, each student will receive his/her own set of questions. Given the recommendation on the redundancy of the question pool (ideally, 1:10), each student will receive a unique test form (Fig. 5).

Then, the test can be made visible in course materials by means of the "eye" icon in the editing mode in course materials or by switching to test settings and enabling the respective option ("Settings" block, "Test Control" group, "Edit Settings"). It should be noted that after the first attempt of test completion, students cannot change the test structure; otherwise, it affects the training statistics. To change the test structure, it is necessary to delete all attempts of test completion and, accordingly, all statistical data. Therefore, the test structure is to be thoroughly worked out so that to avoid further updating [5].

Fig. 5. The off-the-shelf test system

5   Conclusions

Thus, LMS MOODLE system incorporates the extensive functionality for academic performance rating in the form of testing as was shown when developing the test system in the discipline of "Data bases". The element has a flexible configuration enabling the extensive use thereof in electronic smart training: from control testing to training tests and homework. The knowledge control can be carried out both remotely and face-to-face [6].

References

1. Official Site of MOODLE Community (2017). http://www.moodle.org
2. Ukhov, P.A., Lomakin, A.L.: Distant Learning Technology in College: Monography. Moscow Humanistic Technical Academy, 180 p. Publishing House of MHTA (2010)
3. Andreev, A.V., Andreeva, C.V., Dotsenko, I.B.: Practice of E-learning Using MOODLE, 146 p. Publishing House TTI SFD, Taganrog (2008)
4. Khoroshko, L.L., Sukhova, T.S.: Application of computer aided design (CAD) systems for development of electronic educational courses for engineering disciplines in engineering higher educational institution. In: Proceedings of IEEE Global Engineering Education Conference, EDUCON, Germany, Berlin, pp. 644–647 (2013). ISBN 978-1-4673-6109-5
5. Pastuscha, T.N., Sokolov, S.S., Ryabova, A.A.: Creating E-learning Course. Lection in SDL MOODLE: Teaching Aid, 44 p. SPSUWC, SPb. (2012)
6. Khoroshko, L.L., Vikulin, M.A., Kvashnin, V.M.: Technologies for the development of interactive training courses through the example of LMS MOODLE. In: Smart Innovation, Systems and Technologies, pp. 302–309. Springer, Cham (2018). ISBN 978-1-4673-6110-1

New Vision in Microelectronics Education: Smart e-Learning and Know-How, A Complementary Approach

Olivier Bonnaud[1,2(✉)]

[1] GIP-CNFM, Minatec, 38016 Grenoble, France
Olivier.bonnaud@univ-rennes1.fr
[2] Sensor and Microelectronics Department, IETR, University of Rennes 1,
Rennes, France

Abstract. The evolution towards a digital society is indisputable and the development of online tools adapted to education is becoming a priority. At the same time, the technologies that allow the development of these tools as well as the connected objects become more and more complex with a need for know-how as well for the design of these objects as for their manufacture and production. Let us mention that the connected objects cover a large spectrum of applications opened to other disciplines. A new approach to education is needed. Following a presentation of the context, this article discusses the strategy that has been adopted to successfully combine engineering learning, and more especially the microelectronics training with the e-learning and the acquisition of know-how through practice. Because practical training is becoming more and more complex and costly, the trend is to share platforms dedicated to education within the framework of a network. This was set up in France for more than thirty years. The evolution of this network in this new context of digital society with an increasing priority of development of Internet of Things is presented and discussed.

1 Introduction of the e-Learning Based Education

The unbelievable improvement of the microelectronics field corresponding to a higher and higher integration since more than sixty years [1, 2], associated to the huge development of the computer science and computer engineering, and of the incredible increase of communication capabilities with new transmissions techniques and protocols, have led to the development of on-line tools. These tools give access to data centres, for classical information, but also for pedagogical tools such as massive open On-line Courses or MOOCs [3, 4]. Thus, a family of tools has been developed for which the access can depend of the environment academic, industrial, and more general [4]. Thanks to the improvement of the technologies, these tools can contain texts, figures, graphs, but also animations, videos, quiz, and dedicated applets allowing simulations when the associated software are not too much big. Figure 1 shows the

© Springer International Publishing AG, part of Springer Nature 2019
V. L. Uskov et al. (Eds.): KES SEEL-18 2018, SIST 99, pp. 267–275, 2019.
https://doi.org/10.1007/978-3-319-92363-5_25

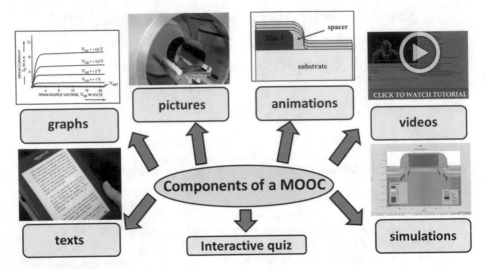

Fig. 1. Different components of an advanced MOOC. In the up-to-date versions, they can contain, videos, simulations, animations and interactive quiz.

different components of an advanced MOOC, more especially in microelectronics domain [5]. These tools also required complementary pedagogical activities to better understand the theoretical background, taking into account students' behavior change and of the new content of their knowledge.

This is the origin of the creation of flipped classes [6], serious games-based learning [7], proactive and intelligent classes, all of which attempt to increase student interest and participation in the learning process. While these approaches already tested in many disciplines contribute to the improvement of scientific background acquisition, the main disadvantage of these approaches is related to the lack of know-how and practice in the curricula of students mainly for engineers.

This paper deals with the importance of practical training on real objects, and learning through projects on real objects that provide the mandatory complement of e-learning. The problem of engineering, and more particularly of microelectronics and nanotechnologies, is the increasing cost of platforms allowing up-to-date experiments compatible with industrial needs. The solution is to share the equipment and facilities of the platforms that involve industrial tools (software and hardware) to minimize their purchase and operating expenses. Within the framework of a national network coordinating twelve interuniversity centers, the article describes the permanent adaptation of practical training, the development of innovative student behaviors and concrete results on the improvement of knowledge whatever the level of studies.

2 Knowledge and Know-How in Engineering and Microelectronics

The practice training becomes a major challenge in the formation of engineers, due to:

- the increasing complexity of the technologies of fabrication [8],
- the increasing importance of the simulation that allows a much faster designing of innovative objects [9],
- the enlargement of the application domains of the connected objects that need in many cases a multidisciplinary approach [10],
- the mandatory connection between the theoretical knowledge and the real objects that are created and used in their specific environment [11].

A new approach must be developed in order to adapt the pedagogical approach to these new context. The students must have a solid knowledge that constitutes the background of the discipline, the microelectronics. This knowledge can be acquired by the students thanks to a judicious balance between online courses that are available for everybody and whatever the time and the duration of using, and by associated exercises and training that can be spent through flipped classes or intelligent classes, and serious games-based learning, for example. The online courses can also be devoted to prepare students to the acquisition of know-how thanks to the introduction of technical and technological approaches that contain description tools such as photos, video, animations and simulations. By combining the theoretical and technical online tools it is possible to provide theoretical knowledge and prepare students for practical training. Practice on dedicated platforms can complement understanding and bring the know-how. In the field of microelectronics, the dedicated online tools can concern all the compartments of the discipline from the design, the fabrication, and the characterization of the objects. But due to the diversification of the applications, online tools can be developed in many other disciplines, the final aim being the creation and the effective realization of connected objects.

3 Organization of the Practice

3.1 Specificity of the Microelectronics Field

In the field of microelectronics, this practice includes several main aspects. The first one is the computer aided design (CAD) of circuits and systems, which is also composed of high level description languages like VHDL, functional modelling, design and simulation of electronics devices and circuits, and generation of layout allowing the creation of masks for the fabrication of the integrated circuits [9]. More recently, new CAD versions allow also the description and the simulation of package corresponding to assembling many circuits on assembled boards. It is included in what is called Systems in Package or SIP. These sets of software are huge and need data base or libraries of hundreds gigabytes. They need powerful computing machine and servers in order to minimize the computation time.

The second aspects is the fabrication technologies. They moved from millimetres to nanometres scales during the last sixty years [1, 2]. The equipment are becoming of an extreme precision and must be protected against environmental contamination, chemical and physical aggressions, with a control of the dimension at the atomic level. All the fabrications are proceeded in clean-room that are strongly protected from dusts, and chemical reactants and strictly controlled in pressure, temperature and humidity. The equipment and associated facilities, even dedicated to the formation and not for the production are becoming very expensive.

The domain of applications of the circuits and devices is wider and wider. The equipment are thus much more diversified that correspond also to a multiplication of the platforms and to an increasing invest dedicated to education.

3.2   Opening Towards Application Domains

As already presented in previous papers [12], the fields of application is wider and wider. Let us to mention the increasingly importance internet of things and connected objects that are involved today in many compartments of the societal needs:

- health with medical assistance and hospital equipment,
- environment with remote control of pollution, contamination and purification,
- energy with monitoring of production and consumption,
- transport with management of mobility and safety of equipment and people,
- communications with all transmission, emission and reception systems,
- information technologies, including cybernetics and computers,

by mentioning only the most known domains.
Figure 2 shows schematically these fields of societal needs.

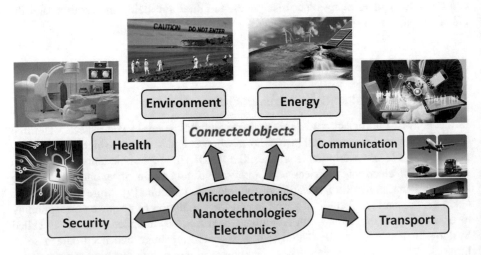

Fig. 2. Diversity of application fields of the connected objects.

3.3   Technical Contents of Connected Objects

Connected objects are thus based on microelectronic design and microelectronic technologies with links to application domains. The main components of a connected object are shown in Fig. 3 [5]. All components that make up connected objects involve devices, circuits or microelectronic systems. Sensors and actuators are the interface with application areas, for example physical, chemical and biological sensors that are generally manufactured in the context of microtechnologies compatible with microelectronic processes [13]. These elements are major in the development of connected objects.

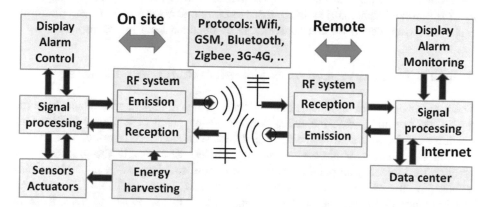

Fig. 3. The different components of a connected object. All these components involve microelectronics devices or systems. The sensors and actuators are the interface with the application fields.

3.4   Practice Associated to Connected Objects

It is clear that future technicians, engineers or physicians must be able to design and manufacture all components of connected objects, which implies a broad spectrum of knowledge and know-how. This is possible on dedicated and always specialized platforms. The equipment are becoming so diversified and so expensive that the only solution is to share the creation and the maintenance of platforms between several institutions within the framework of common centers for education and training. Due to the need for specialization, in practice, there are several centers that cover the entire spectrum of practice in an official network presented below.

4   Organization of Common Platforms. The French National Network

The French national network for education in microelectronics and nanotechnologies (GIP-CNFM) [14, 15], is organized through an official structure, a Public Interest Group, recognized by the Ministry of Higher Education and composed of 14 partners. Twelve are academic institutions in charge of the twelve joint microelectronics centers with other local institutions, and two are industrial organizations, representing the French electronics industry. The twelve academic institutions are the official partners of the GIP structure. This organization allows the development of technological platforms: CAD, cleanroom technological manufacturing, physical, electronic and electrical characterizations, analog and digital architectures, functional tests. These subdomains cover: - the spectrum of powers from low consumption circuits to high power systems, - the spectrum of frequencies, from DC circuits to VHF electronics, - the spectrum of multidisciplinary applications such as mechatronics, optoelectronics, bioelectronics, and the related the sensors and the actuators. Thus, 81 platforms are spread in the 12 joint centres that are opened in 2018 to 89 academic institutions as well as to 60 research laboratories. More than 14,000 students are users of the platforms each year. More details on this network can be obtained from its website [14].

5   Main Innovative Activities Complementary to e-Learning

The number of e-learning tools is exponentially growing, thus many students have the possibility to obtain some theoretical knowledge by using MOOCs, or SPOCs [16] available on-line as well as some dedicated sites such as the "Wikipedia" French data base. A low amount of these tools are nevertheless dedicated to the preparation of practice by including animations, videos or simple simulations. In microelectronics technology, the MOOC created twenty years ago is always available [17] but with some problem of maintenance and up-dating, the tools that were used to create it are no more available today! Let us note that this problem must be taken into account with the globalization of e-learning. In parallel, the network has set-up many innovative platforms able to cover the spectrum of the needs. Figure 4 shows several examples of objects designed, fabricated and characterized on these platforms by users of several common centers.

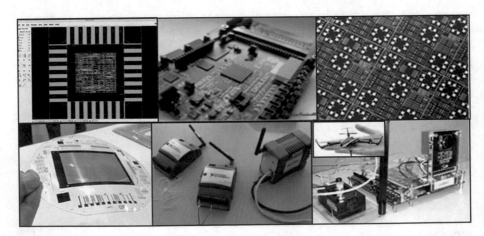

Fig. 4. Innovative practice of students in the joint microelectronics centers. The students design, fabricate and characterize embedded circuits, MEMS, flexible circuits, and connected objects such as energy controllers [18] or drones [19].

6 Discussion and Conclusion

Within the strategy to develop innovating practice adapted to the new pedagogical environment and applied since 2012, more than 60 new practice training subjects are today set-up in the twelve common centers of the CNFM network. During the 2016–2017 academic year, more than 4,000 students had practice on the innovating platforms, representing a total of about 350,000 student-hours. This innovative practice adapted to the arrival of internet of things provides the mandatory know-how to the users and represents about 30% of the total activity of the twelve common centers. All the elements that make up connected objects are addressed thanks to the organization in network. This represents a strength of the French network. The challenge today is to better prepare students to acquire this know-how with the help of specific tools online. These tools should be organized to describe the practical training activity in order to highlight the potential problems and physical limitations that can be verified on the platforms. This will be a way of minimizing the duration of an expensive practice while maintaining a good efficiency of learning the know-how and opening the platforms to more students.

The described approach has been applied to the field of microelectronics which is also a domain at the heart of many fields of application. But it's not limited to this area. Indeed, this approach can be duplicated in many other fields of engineering such as mechanics, civil engineering, robotics, chemistry and energetics, because of the similar tendency of the theoretical approach (increasing of e-learning) and the real requirement of engineer's know-how. In these engineering specialties, more and more expensive and complex platforms are developed. They need to share the facilities dedicated to higher education.

In addition, either domains not especially considered in engineering are also concerned. For example, the domains of the health and the biology are more and more linked to the very precised analysis techniques, process controls, real time treatment of data and real time visualization in order to set-up assisted medical surgeries, for example. All these techniques are involving the connected objects. These objects can also insure the remote control of the people and of their medical environment.

If the context of digital society upcoming, the challenge for Higher Education is high and a new organization of the studies should be set-up. The know-how will be a significant and indisputable strength of the graduate students, technicians and engineers. A lifelong learning for the employees of many companies could be also developed to maintain a high skill of their collaborators able to work in a multidisciplinary and connected environment.

Acknowledgment. The author wants to thank all the members of the French GIP-CNFM network for they contribution to many innovative realizations. This work is financially supported by French Higher Education Ministry and by IDEFI-FINMINA program (ANR-11-IDFI-0017). A special thanks to L. Chagoya-Garzon, secretary of GIP-CNFM for her fruitful advice for the proof reading of this paper.

References

1. Moore, G.E.: Cramming more components onto integrated circuits. Electron. Mag. **38**(8), 114–117 (1965)
2. Simonite, T.: Moore's Law Is Dead. Now What?". MIT Technology Review, 13 May 2016. https://www.technologyreview.com/s/601441/moores-law-is-dead-now-what/
3. Stuchlikova, L., Kosa, A., Benko, P., Donoval, D.: Massive open online courses in microelectronics education. In: Proceedings of 10th EWME 2014, Tallinn, Estonia, 14–16 May, pp. 31–36 (2014)
4. Fox, A.: From MOOCs to SPOCs. In: Proceedings of ACM, vol. 56, no. 12, pp. 38–40 (2013)
5. Bonnaud, O., Fesquet, L.: MOOC and the practice in electrical and information engineering: complementary approaches. In: Proceedings of IEEE ITHET Conference, Istanbul, Turkey, 8–10 September 2016. https://doi.org/10.1109/ithet.2016.7760732
6. Bonnaud, O., Danto, Y., Kuang, Y., Yuan, L.: International flipped class for Chinese Honors bachelor students in the frame of multidisciplinary fields: reliability and microelectronics. In: Proceedings of ICATI 2017 Conference, Samui Island, Thailand, 25 June, p. 12 (2017)
7. Bergeron, B.: Developing Serious Games (Game Development). Delmar Thomson Learning, Boston (2005)
8. Matheron, G.: Keynote, Microelectronics evolution, European, Microelectronics Summit, Paris, France, November 2014
9. Bonnaud, O., Fesquet, L.: A prospective on education of new generations of devices in the FDSOI and FinFET technologies: from the technological process to the circuit design specifications. In: Proceedings of SBMicro 2014, Aracaju-Sergipe, Brazil, pp. 1–4 (2014). https://doi.org/10.1109/sbmicro.2014.6940081
10. Bonnaud, O., Fesquet, L.: Towards multidisciplinarity for microelectronics education: a strategy of the French national network. In: Proceedings of IEEE Microelectronics System Education (MSE) conference, Pittsburg, PA, USA, pp. 1–4. IEEE (2015). ISBN 978-1-4799-9915-6/15/$31.00 ©2015
11. Bonnaud, O., Fesquet, L.: Practice in microelectronics education as a mandatory complement to the future numeric-based pedagogy: a strategy of the French national network. In: Proceedings of EWME, Southampton, UK, May 2016, pp. 1–8. IEEE (2016). https://doi.org/10.1109/EWME.2016.7496460
12. Bonnaud, O., Fesquet, L.: Multidisciplinary topics for the innovative education in microelectronics and its applications. In: Proceedings of 15th International Conference on Information Technology Based Higher Education and Training, Lisboa, Portugal, pp. 1–5 (2015). https://doi.org/10.1109/ITHET.2015.7217961
13. Bonnaud, O.: New approach for sensors and connecting objects involving microelectronic multidisciplinarity for a wide spectrum of applications. Int. J. Plasma Environ. Sci. Technol. **10**(2), 115–120 (2016)
14. CNFM: Coordination Nationale pour la formation en Microélectronique et nanotechnologies (CNFM). www.cnfm.fr; GIP-CNFM: Public Interest Group, administrative structure of the CNFM. Accessed Jan 2018
15. Bonnaud, O., Gentil, P., et al.: GIP-CNFM: a French education network moving from microelectronics to nanotechnologies. In: Proceedings of EDUCON 2011, Amman, Jordan 3–6 April 2011, pp. 122–127 (2011). ISBN 978-1-61284-641-5
16. Garlock, S.: Is Small Beautiful? Online education looks beyond the MOOC. Harvard Magazine, July–Aug 2015. http://harvardmagazine.com/2015/07/is-small-beautiful. Accessed Jan 2018

17. For example Bonnaud, O.: https://microelectronique.univ-rennes1.fr/fr/index.html. Accessed Jan 2018
18. Meillère, S., Ferrero, F., Pannier, P., Jacquemod, G.: Réseaux de Capteurs Intelligents RECAIN/WSN. In: Proceedings of JPCNFM 2014, Saint-Malo, 19 November 2014
19. Fesquet, L., Morin-Allory, K., Rolland-Girod, R.: Un projet de microélectronique numérique original: Contrôle autonome d'un micro-drône par caméras externes. J. J3EA **14** (2015). https://doi.org/10.1051/j3ea/2015021

Smart Platform for e-Learning Circumstances

Dumitru Dan Burdescu(✉)

Department of Computers and Information Technology, University of Craiova,
Craiova, Romania
dburdescu@yahoo.com

Abstract. Mobile learning is a very wide term which encompasses not only learning principles, but also in most cases a bunch of smart devices, from smart-phone to tablet or wearable devices. Any device that allows learning anywhere and anytime is a channel to information and channel that helps gain knowledge and apply it, and this is the main goal of mobile learning. It is not so simple to develop such a system in practice. One of the main objectives of this platform is to adapt students' knowledge and communication skills with the needs of the current economy and society. Another objective why we choose to create disciplines in mobile learning or blended learning format refers to the integration of transversal competences, including the communication in learning and teaching. This paper presents the development of new tools for our smart platform, with implications and analysis from the point of view human computer interaction and mobile learning. The goal of tools is to improve the educational process by helping the professor form a correct mental model of each student's performance. Besides this, the developed applications are also analyzed from human computer interaction and attempt to offer effective and highly usable tools. The languages supported by this module are both English and Romanian.

Keywords: Smart learning process · e-Learning · Mobile learning
Distance learning · Machine learning · Human computer interaction

1 Introduction and Related Works

Smart and collaborative platform is organization of flexible learning solutions and answered to our needs. Our smart platform [1] is organization of flexible learning solutions and one of the main objectives of this platform is to adapt students' knowledge and communication skills with the needs of the current economy and society [6].

Many studies are based on the evaluation of the digital content for the smart system and the importance of the principal-agent interaction in terms of acquiring new competencies is neglected [7, 8]. The principal-agent relationship in the smart system leads to a new challenge: - to use Smart Technology to achieve a motivational learning climate and to develop, at the same time, an individualized and group learning activities. Meanwhile, the principal is able to provide a positive individualized feedback only if he/she transforms the platform in a "friend". Our platform has continuously improved and its facilities generate challenges for the students and the teachers [1, 2, 6, 9]. Therefore, our quantitative and qualitative research put into value the relationship between user and technology.

© Springer International Publishing AG, part of Springer Nature 2019
V. L. Uskov et al. (Eds.): KES SEEL-18 2018, SIST 99, pp. 276–285, 2019.
https://doi.org/10.1007/978-3-319-92363-5_26

In our research, two theoretical approaches are used to develop the hypotheses related to the relationships between user and technology. First, the relationship user - prosumer is used as a basis for a model that predicts the role of the platform Tesys as a tool to promote innovation and entrepreneurial skills. We argued that the platform Tesys is a mechanism used for managing learning process characterized by high levels of interdependence and interconnection. Second, the relationship principal-agent will be sometimes relayed on teachers' level of expertise and sometimes relayed on user/students' competencies. The results indicate that the platform has limited feature for developing entrepreneurial skills to users [10]. Therefore, it is necessary to use additional technology for developing and improving entrepreneurial skills. So, in most cases a bunch of smart devices, from smart-phone to tablet or wearable devices. Any device that allows learning anywhere and anytime is a channel to information and channel that helps gain knowledge and apply it, and this is the main goal of mobile learning.

As illustrated in Fig. 1 above, there are six properties or attributes of our Smart Platform - Tesys [11]. It is designed to provide a cost-effective, stress-free and flexible assessment for the whole practicum experience; aiding the management, supervisors and supervisees. It is also equipped with properties that allow accessibility and reflexivity for both supervisors and supervisees. Further, the whole exercise of doing practicum and assessment online exposes the supervisees to 21st century teaching and learning skills. Concomitantly, the supervisees become more advanced in their technological skills. Stress-free as supervisees are able to perform their teaching in a low-anxiety environment as there is no supervisor sitting at the back of the class observing and making written comments as they are teaching and handling the class. Moreover, the recordings also capture the students' natural behavior in class. As for the supervisors, they are able to evaluate their supervisees' performance at any time that is conducive for grading purposes. They do not need to be physically present in class. Technological-enhancement attribute refers to how the supervisees who are digital immigrants have become more advanced in their technological skills. This advancement or enhancement is important as the supervisees face their own students who are digital natives. They must be relevant to their era, thus the need to keep abreast with the technology today as it plays a big role in tandem with the 21st century teaching and learning skills and the 4.0 Industrial Revolution where the focus is on the internet of things. A reflective exercise involves looking back at one's own current knowledge and past experience. It can be optimized with the help of an expert. The supervisor is the expert who can guide the supervisee to discuss pertinent issues in the classroom management, delivery of lessons and also point out the strengths and weaknesses in the eLearning activities. Thus, this reflective practice (or reflexivity attribute) is conducted via an online forum where both supervisee and supervisor are engaged in an ongoing interaction using to get or give feedback.

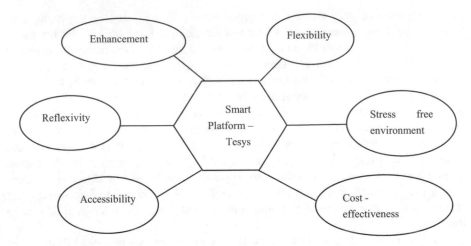

Fig. 1. Properties of our smart platform – Tesys

2 Architecture of the Platform

Nowadays the use a smart platform is a requirement for every university that wants to align to the existing trends and to have a high quality of educational process. Our smart platform [1] is organization of flexible learning solutions and answered to our needs because this platform is used in our university. Custom knowledge representation will enable designing context-aware environments and therefore create the premises for shifting towards intelligent e-Learning environments. The benefits of such approach regard helping professors to prepare courses and high quality e-Learning content for students and life learning users. Learners will benefit by being able to access needed learning material such that their knowledge level will optimally increase and will be adapted. Here I shall present the architecture of the Tesys software system that will manage and use the custom knowledge representation model [2].

The main components of the system are the following:

Central Business Logic Module – this module contains the logic for accessing the e-Learning infrastructure and for sending queries and receiving responses from the Knowledge Miner Module and E-Learning Infrastructure Module. It also represents the main entry point into the software system and human computer interaction research. One of the main objectives of this platform is to adapt students' knowledge and communication skills with the needs of the current economy and society.

Experience Repository Module – this module gathers in a structured format all the data regarding the actions that were performed by learners. The research of human computer interaction and cognitive psychology issues are cornerstone in shifting gears from "technology that solves problems" towards "design that emphasizes the user's needs". These general research areas have a great impact on the field of e-Learning due to the wide range media that can produce cognitive affection at various industries. Among the most common options there are simple text, voice, picture, video, or virtual reality.

Knowledge Model Repository – this module manages the current knowledge model representation. This model is used whenever the intelligent character of an action is needed. That is why the general fundamental usability evaluation formulated by Nielsen [3] needs a proper specific adjustment for e-Learning environments. This module have greater functionality, giving teachers the opportunity to provide contextualized and personalized learning resources, unrestricted by spatial and temporal constraints and allowing learners to use readily-available, handheld computing devices and communication technology to access information from anywhere and at anytime.

Knowledge Miner Module - this module gathers the business logic for querying the knowledge model repository. Many research hours have been allocated to the purpose of extracting key concepts from course materials, messages, questions and finding ways of using them for enhancing the teaching and learning processes. Also, a considerable amount of work has been put into discovering the similarity between concepts. The Natural Language Processing is another major research area, with a strong focus on documents (text and diagrams). In [4] author presents the construction of an English-Romanian tree-bank, a bilingual parallel corpora with syntactic tree-based annotation on both sides, also called a parallel tree-bank. Tree-banks can be used to train or test parsers, syntax-based machine translation systems, and other statistically based natural language applications.

E-Learning Infrastructure Module – this module represents the classical view of an e-Learning environment. The proper operation of Central Business Logic module and Experience Repository Model module is driven by an experience properties file. This file contains the definitions of the actions that are to be logged as experience during the operation of the system. Knowledge Model Repository functionality is managed by a properties file which specifies the employed technique for building the model. This properties file has as input the properties file that sets up the experience repository module.

Knowledge Miner module runs according with the specifications set up by the data analyst. The specifications regard the specific educational goals required by the administrators of the e-Learning environment. Smart infrastructure represents the classical view of a smart platform. It gathers all the assets managed by e-Learning environment: users (e.g. learners, professors, and administrators), disciplines, chapters, course documents, quizzes. It also embeds the needed functionalities for proper running, like security, course downloading, communication, testing or examination. Central Business Logic module along with smart infrastructure represents the classical structure of an e-Learning environment. Experience Repository module, Knowledge Model Repository and Knowledge Miner modules may be regarded as an intelligent component that runs along the e-Learning environment in order to enhance it.

The most important premise for the development a tool is an online educational collaborative platform that has a proper structure for the educational assets and the ability to integrate proper intelligent data analysis techniques [5]. Of course Tesys smart platform may have the following types of components as training modules, content and student modules, assessment subsystem, content and users' modules, data management module and others.

This architecture of the platform allows development of the smart application using MVC architecture. In software system are used the main software components from the

280   D. D. Burdescu

MVC point of view. MainServlet, Action, Manager, Bean, Helper and all Java classes represent the Controller. The Model is represented by the DBMS itself while the Web-macro templates represent the View.

Tesys smart platform is based on well known platforms on market but is more flexible and answered to our needs because this platform is used in faculties of our university. Figure 2 illustrates part of the interface available to the professor for managing the concepts [2]. It is very straight forward, providing the teacher with the list of extracted concepts and some additional options for managing them. These options include: the possibility to add new concepts, modify the existing ones in case they were not correctly extracted and delete the irrelevant concepts, if any.

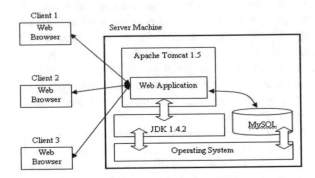

Fig. 2. TESYS – refined software architecture of the platform

Among Services we may enumerate forum, e-mail with staff and others, concept extraction the knowledge, monitoring and recommender modules, mobile learning modules etc. The functionalities of the other modules are affected by their names. The main purpose of smart platform - Tesys for Distance Learning and Mobile Learning is to efficiently manage the learning objects in order to reuse them for creating new educational materials. Sometimes the functionalities of these systems are overlapping but use together in an integrated way lead to providing a complete system solution. Of course smart platform Tesys includes applications for personalization and interaction. Personalization means the learners are in the center of education activity and must provide them flexibility in terms of educational route, time and space of their steps of work. Interaction means facilities of communications between them and to other staff for educational materials and people who take part to educational process.

Starting from the course documents that were previously uploaded by the professor or users on the platform, the software platform extracts the concepts, using a custom concept extraction module, which incorporates a stemming algorithm and formulas. The obtained data is then transferred into the XML files. The five most relevant concepts are also inserted into the Tesys database, for further use. As soon as the professor uploads the test questions and specifies each concept's weight for every question, the student's activity monitoring process can begin. Afterwards, using the concept-weight association, student's responses to the test questions and taking into consideration the performances of learner's colleagues, the software system will be able to show relevant statistics to the professor, so he can understand each learner's learning difficulties as well as the general level of the class.

The tools are designed to review the difficulty of the proposed exam questions and advise the professor on lowering or increasing the exam difficulty. All this process is supervised by the professor, who takes the final decision. Many issues appear when applications contain a mixture of data access code, business logic code, and presentation code. Such applications are difficult to maintain, because interdependencies between all of the components cause strong ripple effects whenever a change is made anywhere. The Model-View-Controller (MVC) design pattern solves these problems by decoupling data access, business logic, and data presentation and user interaction. This architecture of the software platform allows development of the smart application using MVC architecture. This three-tier model makes the software development process a little more complicated but the advantages of having a web application that produces web pages in a dynamic manner is a worthy accomplishment. The model is represented by DBMS (Data Base Management System) that in our case is represented by MySQL. The controller, which represents the business logic of the Tesys platform is Java based, being build around Java Servlet Technology. As servlet container Apache Tomcat 5.0 is used.

3   Smart Tools

One of the main objectives of the our faculties is to facilitate the integration of our students in the national and international labor markets by equipping them with the latest theoretical and applied investigation tools for the IT area. The main objective is to create a transparent set of multidisciplinary courses, seminars and online practical exercises which give students the opportunity to gain both theoretical knowledge and practical skills as well as to develop their key competences. Consistently with computers and initial services provided through the Internet two decades ago, latest development in mobile devices brought changes into the 'traditional' process of ICT implementation into education. Currently, smart and knowledge technologies are of increasing importance [14].

For example, the description of the IT systems implementer position includes following tasks:

- provides concrete solutions to students (mainly abroad), either remote, or on the place;
- independently secures expert support to internal and external students;
- proposes configuration of appropriate solution reflecting requirements pre-defined by the students;
- installs the product which meets students' requirements;
- sets the system so that it met the students' requirements;
- trains students' staff;
- makes documentation of installations and configurations;
- general knowledge of IT environment, good knowledge of office SW, SQL databases, experience in enterprise IS.

The whole program is supported by e-learning lessons. Each module in the e-learning course is equipped with its own guiding instructions, study materials and modified tests with comprehensive sets of questions [2]. Study materials were developed not only as standard presentations; they contain audio-visual materials, animations and instructional video-recordings [5]. Students' attendance in the courses and lectures is monitored in order to evaluate not only the results of the individual tests but also to detect the fields of students' interest [8]. Since smart phones or tablets were launched, the mobile market has rapidly changed. As a result, the field of education is concerned with delivery of knowledge through smart devices (Smart Education). Smart Education mechanism can be seen as an integrated educational environment in which cooperative, interactive, participative, sharing, and intelligent learning are available through new forms of teaching learning content, environment, and ICT (Fig. 3).

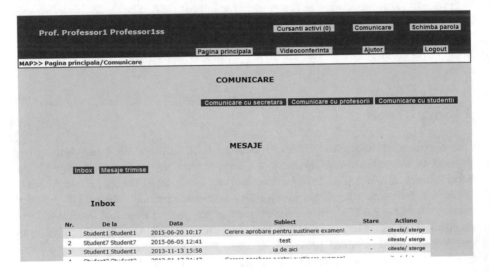

Fig. 3. Main page for teacher's communication

Information and communication technologies have penetrated all spheres of human lives, including education [13]. Reflecting the latest development, mobile devices and technologies are currently under our focus. A shift has been detected from original, i.e. 'immobile' devices and relating technologies to the mobile ones. The current mobile-assisted process of instruction is characterized by specific features. This feature directly predetermines them for educational purposes, starting from the pre-primary to lifelong education, despite there definitely are some didactic constrains and technical limitations. With a mobile learning device the learning process can be rather easily conducted anytime and anywhere. Learning through the computer, phone, tablets etc. enable the learners to learn in a non-classroom environment, e.g. at home; however, learning through the mobile device provides them with the opportunity to learn beyond these classroom and non-classroom environments, i.e. they can genuinely learn every time and everywhere they are [15]. Moreover, the widespread influence of the market increased the popularity of the mobile phone or tablet and this fulfills the need of teachers/designers to provide tools and educational software for the learners in teaching/learning contexts.

Two main characteristics of mobile devices are portability, i.e. the possibility to move the device, and connectivity. It means, when been designed, the mobile systems must have capability of being connected and communicate with the learning website using the wireless network of the device to access learning material ubiquitously, including short message and e-mail service. Moreover, comparing with other wireless devices such as laptop computers, mobile phones or tablets are rather inexpensive having functions as Internet browsers available in most devices. With such inexpensive devices accessible to even the poorest areas and having the functionalities of e-mail or SMS, it is now possible to transfer information to and from mobile phones or tablets between instructors and learners without any difficulty. Although learning services through mobile devices have some advantages, they also have some constraints as small screen, reading difficulty on such a screen, data storage and multimedia limitations etc., particularly mobile phones of small size are not designed for educational purposes.

That is why they may be difficult for the learners to be used for learning. This is partly due to the initial design of such devices, and the purpose of their use—none of mobile devices were primarily developed for the purpose of education.

To illustrate the above mentioned problem, Stockwell [12] demonstrated that the learners found some learning activities take too long to complete on mobile devices, and consequently, some of them preferred to use their PCs to do the assigned task.

The reason is the exploration is essentially mobile, either it means physical movement or movement through conceptual space to form new knowledge. Then, the conversation is the bridge connecting learning across contexts, whether through a discussion, or a phone call between people in different locations, or by making a written comment which can be read at a different time or place. The technology in these explorations and conversations play the role of a mediator of learning. To sum up, mobile learning can be characterized as processes (both personal and public) of coming to know through exploration and conversation across multiple contexts amongst people and interactive technologies. Thus when designing mobile learning, the main task is to promote enriching conversations between learners and teachers within and across contexts. This objective is then reflected in understanding how to design technologies, media and interactions to support a seamless flow of learning across contexts, and how to integrate mobile technologies within education to enable innovative practices. In other words, general principles for human-computer interaction defined in the interaction design research can be applied on mobile devices. These characteristics have been supplemented by more specific findings from mobile learning projects.

4 Conclusions and Future Works

Smart and collaborative platform is organization of flexible learning solutions and answered to Romanian needs. While there are many significant beneficial aspects of collaborative platforms for Distributed Software Development for e-Learning, our experience clearly shows that these benefits are not easily reachable.

Mobile learning technologies offer teachers and students a more flexible approach to learning. Mobile devices have greater functionality, giving teachers the opportunity to provide contextualized and personalized learning resources, unrestricted by spatial and temporal constraints and allowing learners to use readily-available, handheld computing devices and communication technology to access information from anywhere and at anytime.

The simplest situation is when lower wages are countered with the overhead of higher managerial complexities. The software architecture of smart platform proved effective and efficient for implementing the system. Taking into consideration the complexity of the software architecture which makes intensive usage of different tools and technologies the adopted software development process proved to be a reliable one. The smart platform has also implemented administration capabilities for professors, learners and secretaries [16].

Above all, rules for mobile devices exploitation should be defined, identically to the situation two decades ago, when the process of computer-assisted education started.

In the case of mobile devices, the size of screen should be a feature strongly limiting the process of their implementation into education. The use of mobile phones and other portable devices have shown an impact on how learning takes place in many disciplines and contexts, thus providing the potential for significant change in teaching and learning practices. She emphasizes the mobile-assisted is not a stable concept; therefore its current interpretations need to be made explicit so that to highlight what is distinctive and worthwhile about mobile learning. From the long-term view, didactic principles having been defined and applied in e-learning, it is high time technical constraints of mobile devices were seriously considered and reflected. Then the process will become efficient and motivating for learners.

Once progress is made in these directions, similar smart tools may also be analyzed and modules will be designed and implemented for progress in this domain.

References

1. Burdescu, D.D.: TESYS - an e-Learning system. In: The 9th International Scientific Conference Quality and Efficiency in e-Learning, vol. 3 (2013)
2. Burdescu, D.D.: A Platform for e-Learning, vol. 262, pp. 633–640. IO Press-Frontiers in Artificial Intelligence and Applications (2014). https://doi.org/10.3233/978-1-61499-405-3-633
3. Nielsen, J., Molich, R.: Heuristic evaluation of user interfaces. In: Chew, J.C., Whiteside, J. (eds.) Proceedings of the SIGCHI Conference on Human Factors, Computing Systems (CHI 1990), pp. 249–256. ACM, New York (1990)

4. Colhon, M.: Language engineering for syntactic knowledge transfer. Comput. Sci. Inf. Syst. **9**(3), 1231–1248 (2012)
5. Burdescu, D.D.: E-Learning tools for a software platform. In: Uskov, V., Howlett, R., Jain, L. (eds.) Smart Education and Smart e-Leaning, vol. 41, pp. 231–242. Springer, Cham (2015). https://doi.org/10.1007/978-3-319-1985-0
6. Burlea, A.S., Burdescu, D.D.: The development of the critical thinking as strategy for transforming a traditional university into a Smart University. In: Uskov, V., Howlett, R., Jain, L. (eds.) Smart Education and e-Learning 2017. SEEL 2017. Smart Innovation, Systems and Technologies, vol. 75, pp. 67–74. Springer, Cham (2018). https://doi.org/10.1007/978-3-319-59451-4_7
7. Politis, D.: The process of entrepreneurial learning: a conceptual framework. Entrep. Theor. Pract. **29**(4), 399–423 (2005)
8. Tacu, M.G., Mihaescu, M.C., Burdescu, D.D.: Building professor's mental model of student's activity in on-line educational system. In: The Third International Conference on Cognitonics, (Cognit-2013), Ljubljana, Slovenia (2013)
9. Burdescu, D.D., Mihaescu, C.M.: Building intelligent e-learning systems by activity monitoring and analysis. In: Tsihrintzis, G., Jain, L. (eds.) Multimedia Services in Intelligent Environments. Smart Innovation, Systems and Technologies, vol. 3, pp. 153–174. Springer, Heidelberg (2010)
10. Burlea, A.S.: The complexity of an e-learning system: a paradigm for the human factor. In: The Inter-Networked World: ISD Theory, Practice and Education, vol. 2, pp. 267–278. Springer, Boston (2008)
11. Halawa, A., Sharma, A., Bridson, J.M., Lyon, S., Prescott, D., Guha, A., Taylor, D.: Constructing quality feedback to the students in distance learning: review of the current evidence with reference to the online master degree in transplantation. World J. Educ. **7**(4), 117–121 (2017)
12. Stockwell, G.: Investigating learner preparedness for and usage patterns of mobile learning. ReCALL **20**(3), 253–270 (2008)
13. Schiopoiu, A.S., Remme, J.: The Dangers of Dispersal of Responsibilities. Amfiteatru Econ. **19**(45), 464–476 (2017)
14. Burlea, A.S., Burdescu, D.D.: An integrative approach of e-Learning: from consumer to prosumer. In: Smart Education and e-Learning 2016. Smart Innovation, Systems and Technology, vol. 59, pp. 269–279. Springer International Publishing, Switzerland (2016). https://doi.org/10.1007/978-3-319-39690-3
15. Burlea A.S.: Success factors for an information systems projects team: creating new context. In: 11th IBIMA Conference, 4–6 January 2009, Cairo, Egipt, pp. 936–941 (2009)
16. Burlea, A.S.: The complexity of an e-Learning system: a paradigm for the human factor. In: The Inter-Networked World: ISD Theory, Practice and Education, vol. 2, pp. 267–278. Springer, Boston (2009). https://doi.org/10.1007/978-0-387-78578-3_21

Engineering Affordances for a New Convergent Paradigm of Smart and Sustainable Learning Technologies

Steven Coombs[1]([✉]) [iD] and Madhumita Bhattacharya[2,3] [iD]

[1] Hamdan Bin Mohammed Smart University, Dubai, UAE
s.coombs2015@gmail.com
[2] New Paradigm Solutions Ltd., Palmerston North 4410, New Zealand
mmitab@gmail.com
[3] Global Association for Research and Development of Educational Technology,
Victoria, BC, Canada

Abstract. The proposed new paradigm for smart and sustainable learning technologies has been created from the notion of how and why smart thinking supports smart learning. Indeed, smart thinking has been pedagogically associated with the use of cognitive tools to assist deeper learning in the form of scaffolding in general and e-scaffolding in online environments. From this basis the idea of a smart learning technology has been linked to a sustainable learning technology in terms of its affordances and ability to evolve and converge. This idea has been developed from a sustainable change model built upon self-organizing systems related to useful and necessary evolutionary improvement. Finally, a new framework for smart sustainable learning technologies has been proposed. This paradigm maintains that a smart learning environment is supported by e-scaffolding interactive tools for the co-construction of knowledge, communication and interpersonal skills and reflective thinking skills, along with peer and self-assessment for learning. It can represent a convergent set of sustainable learning technologies such as eportfolios for learning and assessment.

Keywords: Smart thinking · e-Scaffolding · Smart learning technologies
Sustainable learning technologies · Cognitive map · Cognitive tools
Critical thinking scaffolds · Problem based learning · Affordances
ePortfolio · Convergent technology · Connectivism · Peer and self-assessment

1 Smart and Sustainable Learning Technologies

1.1 Smart Learning Technologies

In order to consider what a smart learning technology is the authors wish to first deconstruct the concepts of smart and smart learning. So the question is: What is Smart and what is smart learning? The authors would like to consider one definition of *smart* as being an efficiency and effectiveness model of maximized utilization of all resource inputs, e.g. time invested, money/economic, human resources (physical and mental abilities plus experience), availability of physical resources. This would also imply that smart learning somehow maximizes the effectiveness and efficiency of human learning.

Indeed, in order to understand more fully the nature of smart learning Coombs and Bhattacharya [1] proposed that in fact smart thinking was at the heart of understanding what smart learning really is. They argued that smart thinking was a systematic form of 'scaffolded' critical thinking [2, 3] either through human facilitator intervention or perhaps through the use of digital cognitive scaffolds as a form of *e-scaffolding* [4] to enable learning.

The key argument is that smart learning requires learners to engage in smart thinking and that the pedagogical role and designed nature of such cognitive tools operates as an important support technology for learners. Coombs and Bhattacharya [1, 5] provided an important postulate regarding the relationship between smart learning and thinking and defined the nature of 'smart' in terms of being applied to a sustainable learning environment:

Smart is defined in terms of critical and systematic thinking processes that lead to sustainable forms of continuous learning within a culturally framed learning environment.

The authors therefore suggest the use of critical thinking cognitive tools operating as critical thinking scaffolds to help elicit both individual and peer-based learning conversations in virtual environments that can use techniques such as problem based learning [6] to help seek the need for smart solutions.

A smart learning technology can be supported by a smart thinking and learning environment with the help of suitable cognitive tools, which also help to ensure transfer of knowledge and lifelong learning [1]. In this way a smart learning technology becomes a sustainable learning technology where online cognitive tools assist in the modelling of conceptual knowledge. In addition, a smart learning technology is adaptable to the needs of the learner. Consequently, a smart learning technology can be designed to offer a differentiated and personalized approach towards learning and formative assessment that can be supported with learning analytic data for real-time feedback and feedforward and has been proposed as Formalytics [7].

Smart cognitive tools that are responsive to the learner's real-time needs therefore support a smart learning technology. Such learning technologies are also linked to a Design Based Model (DBM) cognitive approach towards the conceptual modeling of knowledge, associated social skills and one's personal value system. This is achieved through constructivist scaffolding and eliciting of meaningful learning experiences via cognitive digital tools that embed new repertoires and schemas of reflective thinking [3] as evident in eportfolios for learning [8].

Smart learning cognitive tools such as digital cognitive mapping [9, 10] for collaborative group work could be designed for working on focused projects, e.g. product development at a distance, which could then be extended for larger scale implementation to achieve results in much less time [11]. The use of smart learning technologies to facilitate the process of engaging in collaborative knowledge creation also ensures enhanced social and cognitive development [12, 13] and therefore confirms sustainability of the learning technology through greater engagement.

1.2   The Case for Sustainable Change

In our modern age sustainability is considered important, not only in environmental matters but in educational systems as argued by UNESCO's 17 Sustainable Development Goals (SDGs) [14]. The UN's SDG 4 relates to sustainable education with the mission to: *Ensure inclusive and equitable quality education and promote lifelong learning opportunities for all.* So in answer to the question: Who cares about sustainability? The answer lies at the cutting edge agenda for the UN and all its member countries. The definition of sustainability, however, is another matter. The authors argue that sustainability should not be mistaken for continuity as some kind of steady-state system. Instead, sustainability represents a continuity of evolutionary change. Evolutionary change is understood to happen in self-organizational systems for helping to understand the management of enterprises engaged in a constant state of change and restructuring [15]. The self-organizational approach provides a systematic means of modelling evolutionary and revolutionary changes in formal organizations that can also be seen as a form of continuous and sustainable change. Likewise, in the educational world there is a theory of self-organized learning [16] that considers learning to learn by learners to be in a constant state of personal change and continuous growth governed by a systematic form of a learning conversation that models critical reflection and the construction of knowledge [17]. The learning-to-learn models underpinning self-organized learning are considered to be a lifelong and sustainable journey whereby the learner's capacity-to-learn and personal growth is enhanced over time.

The Design Based Evaluation (DBE) approach and subsequent model outlined in [18] represents a sustainable approach towards implementing ongoing projects associated with dynamic change. Thus, the DBE approach [18] represents an evolutionary cycle and model for implementing *sustainable change* over the longer term. Sustainable change can be seen as the engine of evolution built upon a self-organizational system that can model the processes involved. Evolution is generally understood as the survival of the fittest with what survives becoming the sustainable solution over time. The eye developed into a sustainable bio-technology for most animal life on earth with many useful further refinements required for sustaining survival. This principal of sustainable change and the subsequent need for refinements defines the nature of sustainable change. Sustainable change can be defined and modeled from many perspectives. Some critical theorists [19] have contextualized sustainable change as an activities-led intervention model to pursue sustainable wellbeing as a form of actualizing happiness in social and work settings. However, systematic and self-organized activities linked to some defined agenda for change with regular cycles for articulating key learning events, as opportunities for individual and group-based critical reflection [20] appear to define a sustainable change agenda.

2   The Nature of a Sustainable Learning Technology

2.1   Affordances of Sustainable Learning Technologies

Affordances of all systems are important because they are the key aspect of sustainability of anything, including technologies, because we need to identify the characteristics and properties of the entity (physical, psychological – ideas, concepts etc.) concerned and how they can be utilized.

Seminal theorists such as Norman [21] helped to make sense of the conceptual framework of *affordance* in terms of the design aspect of an object related to how the object can be used prompted by some visual clue suggesting its function in terms of perceived properties and use, e.g. visual graphics used to support e-commerce on the internet. Norman [21] stated that:

"...the term affordance refers to the perceived and actual properties of the thing, primarily those fundamental properties that determine just how the thing could possibly be used."

Norman thus defines an affordance as something of both actual and perceived properties and can be further understood in terms of an entity's articulated properties and its potential uses. By understanding the affordances of a learning technology in terms of its evolved improvements over time we can also move to the idea of charting its affordances means of its convergence with other technologies linked to necessary improvements and therefore becoming sustainable.

2.2   Convergence of Affordances and Actions for Sustainability

The convergence of technologies as both systems and connected ideas bring one or more extant technologies and their combined affordances together resulting in a new and expanded technology. Linked to the idea of sustainable change and improvement as the evolution of useful technology we have the convergence of affordances underpinning the nature of sustainability. A good example of a convergent technology is the smart phone and for convergence of learning technologies we might consider the eportfolio [22]. In the eportfolio we have a convergence of many learning and knowledge technologies as described in Fig. 1. In this case a collaborative group working on a project can create both integrated and personal eportfolios. Activities involve discussion, brainstorming, negotiations etc., in order to co-construct knowledge and generate creative ideas. Then the group members work together and individually in co-operation and collaboration through various steps provided in a Design Based Learning Environment using the iterative routine of a Design Based Research methodology [18].

2.3   The Nature of a Sustainable Learning Technology

We now have a model for sustainable change that can also relate to the nature of technological change as an evolutionary process of refinement in order to solve ongoing emerging problems as they arise in real time and also unpredictably. Other theorists from the business world [23] talk about disruptive technology and its associated effect of being a disruptive innovation. However, the issue is that in this approach displaced services and people caught in the wave of new innovations are casualties of any change process due to a competing market rather than one that collaborates, e.g. large sustainable projects like AirBus, NASA etc., require systematic collaboration and are given ongoing government, indeed, inter-governmental support to guarantee their strategic importance and future sustainability. Thus, the need for a collaborative process of

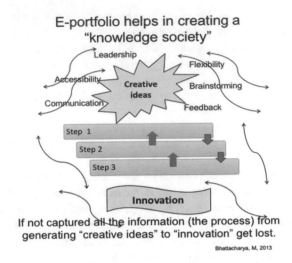

Fig. 1. The convergent nature of an innovative and sustainable learning technology

inclusion within change to mitigate, if not totally avoid, the potential social disruption of the technology. For example Siemens [24] proposes the social agenda of connectivism as a process to support the networking of ideas and knowledge as a continual learning process and Dron and Anderson [25] argues for arbitration and collective wisdom of social participants affected by technological change.

The authors propose that a sustainable learning technology is related to a smart thinking and sustainable learning environment [1, 5] that provides enriched interactive learning through systematic intervention using smart thinking tools to stimulate deeper learning and focused critical reflection. This has been represented by the smart sustainable learning technology schema represented in Fig. 1.

The schema in Fig. 2 depicts how a smart learning environment links the elements of curricula in terms of the learning pathways between learning resources, tasks and assessment. Learners are supported through e-scaffolding via the use of diverse cognitive tools that support the curricula tasks to be achieved. At the same time the e-scaffolding embeds knowledge through critical reflection engagement with both the curriculum represented as a *cognitive map* and iterative feedback and feedforward with an expert as an e-facilitator as well as with other peer learners. This continual learning process supports the ideas of connectivism [25] and self-organized learning [16] and underpins lifelong learning and the sustainable learning agenda in general.

Further, the authors propose a working definition for any sustainable learning technology:

A sustainable learning technology can be represented by a smart learning environment supported by e-scaffolding interactive tools linked to an academic framework operating as a cognitive map of the learning process. Such a cognitive map provides learning pathways as collaborative tools from which e-content is researched, critically reflected upon and modeled into new forms of embedded knowledge for future transfer and articulation.

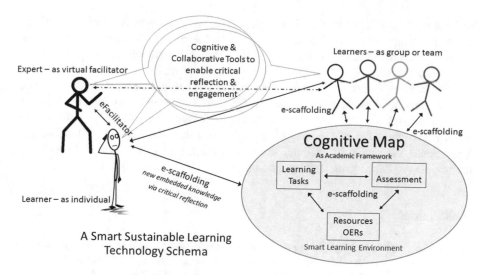

Fig. 2. The smart sustainable learning technology conceptual framework

However, the pedagogical cautions of many so-called smart technologies is that whilst they can be used as a smart learning technology, in practice they may not; since they are not being used in practice to engage the user in smart thinking. A good example is the smart phone that offers great potential to support smart learning environments, but in practice is not being fully utilized. Such ubiquitous devices need to be part of the smart learning design governed by pedagogical protocols that ensures their proper use. Another good example of a smart technology to consider is the eportfolio that has the potential to be a very useful smart learning technology. However, if the eportfolio is only used as a resource-based repository then it does not satisfy the earlier definition of smart, i.e. critical and systematic thinking within a socio-cultural context as a form of rich interactive engagement.

3   Concluding Remarks and Insights

In this research paper the authors have introduced the concept of smart learning technologies linked to sustainable change. The authors have also reviewed the nature of sustainable change in terms of the necessity of improvements of systems evolving over time and the fact that the processes supporting this development are self-organizing and that for a sustainable learning technology can be understood through the learning theory of self-organized learning.

With critical and creative thinking skills embedded into a smart learning environment supported by e-scaffolding cognitive tools and self-organized learning processes the authors have identified the basis of a sustainable learning technology. By understanding the affordances of a learning technology in terms of its evolved improvements of developed properties and uses over time and also integration with other technologies as a form of convergence; the nature of a sustainable learning technology was explained. Combined with e-scaffolding cognitive tools operating as smart thinking tools the intention was to stimulate deeper interactive learning and focused critical reflection. From this basis a smart sustainable learning technology conceptual framework was proposed.

The authors recommend further research and investigation into practical systems that develop the ideas supporting a smart sustainable learning technology environment. Perhaps through the development and testing of bespoke cognitive tools designed to support and elicit the affordances of emerging learning and knowledge technologies. This could be achieved through using a design-based research and evaluation methodology [18] as proposed by the authors.

References

1. Coombs, S., Bhattacharya, M.: Smart thinking smart learning: sustainable learning systems for a sustainable future. In: Anwar., S., Ankit, A., AlZouebi, K. (eds.) Proceedings of Innovation Arabia 10, pp. 68–74. Hamdan Bin Mohammed Smart University, Dubai, United Arab Emirates (2017). https://www.hbmsu.ac.ae/downloads/massmail/2017/august/HBMSU_Innovation_Arabia_Smart_Learning_Conference_Proceedings_2017.pdf
2. Coombs, S.: The psychology of user-friendliness: the use of information technology as a reflective learning medium. Korean J. Think. Probl. Solving **10**(2), 19–31 (2000). Korea: Keimyung University
3. Coombs, S.J.: Design and conversational evaluation of an information technology learning environment based on self-organised learning. Doctoral thesis, Centre for the Study of Human Learning, Brunel University, London (1995). bura.brunel.ac.uk/bitstream/2438/4829/10/FulltextThesis_Volume1.pdf. Accessed 15 Mar 2018
4. Coombs, S.: e-scaffolding: An epistemological framework for e-learning. In: Bastiaens, T., Carliner, S. (eds.) Proceedings of World Conference on E-Learning in Corporate, Government, Healthcare, and Higher Education, pp. 1449–1467. AACE, Chesapeake (2007). http://www.editlib.org/p/26553. Accessed 15 Mar 2018
5. Coombs, S., Bhattacharya, M.: Smart learning requires smart thinking: the evolution of sustainable learning environments. In: Dron, J., Mishra, S. (eds.) Proceedings of E-Learn: World Conference on E-Learning in Corporate, Government, Healthcare, and Higher Education, pp. 303–313. Association for the Advancement of Computing in Education (AACE), Vancouver (2017). https://www.learntechlib.org/p/181201/. Accessed 15 Mar 2018
6. Bhattacharya, M.: Conducting problem based learning online. In: Proceedings of International Conference on Computer in Education, pp. 525–530. RMIT University, Melbourne (2004)
7. Bhattacharya, M., Coombs, S. Formalytics as real-time feedback and feedforward for sustainable lifelong learning pathways. In: Dron, J., Mishra, S. (eds.) Proceedings of E-Learn: World Conference on E-Learning in Corporate, Government, Healthcare, and Higher Education, pp 303–313. Association for the Advancement of Computing in Education (AACE), Vancouver (2017). https://www.learntechlib.org/p/181201/. Accessed 15 Mar 2018
8. Heinrich, E., Bhattacharya, M., Rayudu, R.: Preparation for lifelong learning using ePortfolios. Eur. J. Eng. Educ. **32**(6), 653–663 (2007). https://pdfs.semanticscholar.org/3d0c/7639aced13471a2518de25a35a37be6beb70.pdf. Accessed 15 Mar 2018

9. Bhattacharya, M.: Cognitive maps as a tool for discussion in a computer supported collaborative learning environment. In: International Conference on Cognitive Science, Tokyo, Japan, pp. 326–330 (1999)

10. Bhattacharya, M.: Study of asynchronous and synchronous discussion on cognitive maps in a distributed learning environment. In: Proceedings of WebNet99-World Conference on the WWW and Internet, Hawaii, USA, vol. 1, pp. 100–105 (1999)

11. Koo, B.: Knowledge Management in the Age of Artificial Intelligence: An approach based on Open Source Community Practices, Keynote at Innovation Arabia 11, Dubai, UAE, 11–13 March (2018)

12. Bhattacharya, M., Chatterjee, R.: Collaborative innovation as a process for cognitive development. J. Interact. Learn. Res. **11**(3/4), 295–312 (2000). Special Issue on Intelligent Systems/Tools in Training and Life-long Learning. https://www.learntechlib.org/p/8381/. Accessed 15 Mar 2018

13. Bhattacharya, M., Narita, S.: Design of a computer based constructivist tool for collaborative learning. In: Crawford, C., Davis, N., Price, J., Weber, R., Willis, D. (eds.) Proceedings of SITE 2003–Society for Information Technology and Teacher Education International Conference, pp. 3251–3254. Association for the Advancement of Computing in Education (AACE), Albuquerque (2003). https://www.learntechlib.org/p/18686/. Accessed 15 Mar 2018

14. UN: Transforming our world: the 2030 Agenda for Sustainable Development. United Nations – Sustainable Development knowledge platform (2015). https://sustainable-development.un.org/post2015/transformingourworld. Accessed 2018/03/15

15. Sundarasaradula, D., Hasan, H., Walker, D.S., Tobias, A.M.: Self-organization, evolutionary and revolutionary change in organizations. Strat. Change **14**, 367–380 (2005). https://doi.org/10.1002/jsc.739

16. Thomas, L., Harri-Augstein, S.: Self-Organised Learning: Foundations of a Conversational Science for Psychology. Routledge & Kegan Paul, London (1985)

17. Harri-Augstein, S., Thomas, L.: Learning Conversations: The Self-Organized Learning Way to Personal and Organizational Growth. Routledge & Kegan Paul, London (1991)

18. Bhattacharya, M., Coombs, S.: Proposing an Innovative Design Based Evaluation Model for Smart Sustainable Learning Technologies, KES SEEL-18, Brisbane, Australia, 20–22 June (2018)

19. Lyubomirsky, S., Sheldon, K.M., Schkade, D.: Pursuing happiness: the architecture of sustainable change. Rev. Gen. Psychol. **9**(2), 111–131 (2005). https://escholarship.org/uc/item/4v03h9gv. Accessed 15 Mar 2018

20. Bhattacharya, M.: Introducing integrated e-portfolio across courses in a postgraduate program in distance and online education. In: Spratt, C., Lajbcygier, P. (eds.) E-Learning Technologies and Evidence-Based Assessment Approaches, pp. 243–253. IGI Global, Hershey (2009). https://doi.org/10.4018/978-1-60566-410-1.ch014

21. Norman, D.A.: The Psychology of Everyday Things. Basic Books, New York (1988)

22. Bhattacharya, M.: Convergence of multiple digital technologies for working collaboratively at a distance. In: Keynote at the India-Canada International Conference on Open and Flexible Distance Learning, SNDT Women's University, Mumbai, India, 20–22 February 2013

23. Christensen, C.: Disruptive technologies catching the wave. Harv. Bus. Rev. 3 (1995)

24. Siemens, G.: Connectivism: A learning theory for the digital age (2004). http://www.elearnspace.org/Articles/connectivism.htm last accessed 2018/03/15

25. Dron, J., Anderson, T.: Teaching Crowds: Learning & Social Media. AU Press, Athabasca (2014)

Author Index

© Springer International Publishing AG, part of Springer Nature 2019
V. L. Uskov et al. (Eds.): KES SEEL-18 2018, SIST 99, pp. 295–296, 2019.
https://doi.org/10.1007/978-3-319-92363-5

Printed in the United States
By Bookmasters